77

Advances in Biochemical Engineering/Biotechnology

Managing Editor: T. Scheper

Springer-Verlag Berlin Heidelberg GmbH

Chip Technology

Volume Editor: Jörg Hoheisel

With contributions by
A. Brazma, K. Büssow, C. R. Cantor, F. C. Christians,
G. Chui, R. Diaz, R. Drmanac, S. Drmanac, H. Eickhoff,
K. Fellenberg, S. Hannenhalli, J. Hoheisel, A. Hou,
E. Hubbell, H. Jin, P. Jin, C. Jurinke, Z. Konthur,
H. Köster, S. Kwon, S. Lacy, H. Lehrach, R. Lipshutz,
D. Little, A. Lueking, G. H. McGall, B. Moeur,
E. Nordhoff, L. Nyarsik, P. A. Pevzner, A. Robinson,
U. Sarkans, J. Shafto, M. Sohail, E. M. Southern,
D. Swanson, T. Ukrainczyk, D. van den Boom, J. Vilo,
M. Vingron, G. Walter, C. Xu

 Springer

Advances in Biochemical Engineering/Biotechnology reviews actual trends in modern biotechnology. Its aim is to cover all aspects of this interdisciplinary technology where knowledge, methods and expertise are required for chemistry, biochemistry, microbiology, genetics, chemical engineering and computer science. Special volumes are dedicated to selected topics which focus on new biotechnological products and new processes for their synthesis and purification. They give the state-of-the-art of a topic in a comprehensive way thus being a valuable source for the next 3–5 years. It also discusses new discoveries and applications.

In general, special volumes are edited by well known guest editors. The managing editor and publisher will however always be pleased to receive suggestions and supplementary information. Manuscripts are accepted in English.

In references Advances in Biochemical Engineering/Biotechnology is abbreviated as *Adv Biochem Engin/Biotechnol* as a journal.

Visit the ABE home page at http://link.springer.de/series/abe/
http://link.springer-ny.com/series/abe/

ISSN 0724-6145
ISBN 978-3-662-14644-6   ISBN 978-3-540-45713-8 (eBook)
DOI 10.1007/978-3-540-45713-8

Library of Congress Catalog Card Number 72-152360

http://www.springer.de

© Springer-Verlag Berlin Heidelberg 2002
Originally published by Springer-Verlag Berlin Heidelberg New York in 2002
Softcover reprint of the hardcover 1st edition 2002

Typesetting: Fotosatz-Service Köhler GmbH, Würzburg
Cover: E. Kirchner, Heidelberg

Printed on acid-free paper   SPIN: 10846806   02/3020mh – 5 4 3 2 1 0

Advances in Biochemical Engineering Biotechnology also Available Electronically

For all customers with a standing order for Advances in Biochemical Engineering/Biotechnology we offer the electronic form via LINK free of charge. Please contact your librarian who can receive a password for free access to the full articles. By registration at:

http://www.springer.de/series/abe/reg_form.htm

If you do not have a standard order you can nevertheless browse through the table of contents of the volumes and the abstracts of each article at:

http://link.springer.de/series/abe/
http://link.springer_ny.com/series/abe/

There you will find also information about the

– Editorial Board
– Aims and Scope
– Instructions for Author

Attention all Users
of the "Springer Handbook of Enzymes"

Information on this handbook can be found on the internet at
http://www.springer.de/enzymes/

A complete list of all enzymes entries either as an alphabetical Name Index or as the EC-Number Index is available at the above mentioned URL. You can download and print them free of charge.

A complete list of all synonyms (more than 25,000 entries) used for the enyzmes is available in print form, ISBN 3-540-41830-X.

Save 15%

We recommend a standing order for the series to ensure you automatically receive all volumes and all supplements and save 15% on the list price.

Preface

DNA-chip analysis has come a long way since the first official gathering of scientists working in this field, which took place in Moscow in 1991, comprising about 50 scientists from Europe and the USA. Then, the initial aim was the development of a technology for high-throughput sequencing of the human genome, named sequencing by hybridisation. The field soon diversified, however, in terms of methodology and application. Nowadays, DNA-microarrays seem to be a common commodity in biological sciences. The complexity hidden behind the apparent ease of such studies, however, is highlighted by the fact that it took about ten years of technology development – and persuasion – before the methodology really took off. Also, at closer scrutiny one realises that many problems still need to be resolved and only relatively limited inquiries have been attempted so far considering the potentially possible. Nevertheless, even these experiments have produced data on a scale beyond imagination for most people in biology a few years ago and for many even today. Even the data sets originating from large-scale sequencing are dwarfed by the quantity of information from chip-based analyses. Thus, DNA-microarray technology will be the true herald of change in biology. Having developed earlier from a descriptive to an analytical science, biology will split in future into an experimental and a theoretical branch, similar to what happened in physics quite a long time ago.

This change in biology is no better represented than by the authors of this book, who took part in bringing about this shift in emphasis. They are well-known experts in the field, many – like Edwin Southern, Hans Lehrach, Radoje Drmanac, Pavel Pevzner and Charles Cantor – have been actively pursuing array technology for more than a decade. Rather than pondering on the good old times, however, they demonstrate the continuous development in both technology and application areas and elucidate some critical points that need to be considered, when performing microarray analyses.

The first article, by Pavel Pevzner and co-workers, informs on aspects of designing DNA-arrays, which is not a trivial undertaking, although rarely discussed. Even at this level of array-based analysis – right at the start rather than the end – informatics is required in order to deal with the inherent complexity.

Such a design phase is followed by the actual production process. Although by no means the sole procedure for doing so, photolithographically controlled in situ synthesis is currently the best advanced technique for large-scale chip production with a high degree of reproducibility. Glen McGall and Fred Christians report on the procedures involved, some future developments and applications.

Muhammad Sohail and Edwin Southern describe in their contribution a different process for the creation of oligonucleotide arrays. Subsequently, they used the arrays in the screening for effective antisense reagents. This study is fascinating because of its impact on the understanding of interactions between nucleic acids, an interesting research area even after 50 years of structural DNA analysis, and in view of its practical implications for applications in biotechnology and molecular medicine.

Hubert Köster, Charles Cantor and co-workers combine two high-throughput techniques – arrays with detection by mass spectrometry – for genotyping analyses. With the existence of about 1.5 million publicly known single nucleotide polymorphisms (SNPs), the exploration of this resource is a major challenge in the effort of translating basic genomics into applied, medically relevant molecular genetics.

The work of Radoje Drmanac extends the objective of genotyping analyses and, at the same time, returns to the origin of DNA-chip analysis by pursuing 'sequencing by hybridisation', which is nothing short of performing a SNP analysis for each and every nucleotide of a given DNA-fragment. He reports recent achievements and deliberates on the exciting opportunities of this methodology.

The text of Holger Eickhoff and colleagues already reaches beyond the mere DNA-chip by reporting on procedures that extend strongly into the field of proteomics, thus linking the two areas. Only by such measures, carried out experimentally as well as in silico, the complexity of functions in cellular systems will eventually be unravelled.

Considering how array analyses are performed, it is only natural that a contribution on bioinformatics tools and databases should come at the end of the list. Its position does not reflect its importance, however. As a matter of fact, all preparatory work going into the production of the nice looking, colourful pictures from DNA-arrays is a useless squander unless it is assessed and presented in a way that makes the data accessible to human interpretation. Currently, much of the data produced on microarrays is actually wasted. Transcriptional profiling studies, for example, usually concentrate on few, specific biological aspects and ignore much else contained in the very set of raw data. This information could be useful for other studies, if only one could access it. For the purpose of going back to results for entirely different analytical purposes, central databases with appropriately designed and standardised procedures, as well as a common ontology, are essential. Alvis Brazma and colleagues have been instrumental in getting such efforts started.

Overall, the various articles provide a good mix, covering many, although not all, aspects of microarray-based analysis, the latter no longer achievable in a single book, for the days of the Moscow meeting in 1991 are long past and the breadth of the field has expanded enormously both in terms of its technical aspects and the variety of potential applications. Nevertheless, I hope the picture is comprehensive enough for understanding the basics, elaborate enough to inform in detail on certain aspects and speculative enough to stimulate further developments.

Heidelberg, May 2002   Jörg D. Hoheisel

Contents

Combinatorial Algorithms for Design of DNA Arrays
S. Hannenhalli, E. Hubbell, R. Lipshutz, P. A. Pevzner 1

High-Density GeneChip Oligonucleotide Probe Arrays
G. H. McGall, F. C. Christians 21

Oligonucleotide Scanning Arrays: Application to High-Throughput
Screening for Effective Antisense Reagents and the Study of Nucleic Acid
Interactions
M. Sohail, E. M. Southern 43

The Use of MassARRAY Technology for High Throughput Genotyping
C. Jurinke, D. van den Boom, C. R. Cantor, H. Köster 57

Sequencing by Hybridization (SBH): Advantages, Achievements,
and Opportunities
R. Drmanac, S. Drmanac, G. Chui, R. Diaz, A. Hou, H. Jin, P. Jin, S. Kwon,
S. Lacy, B. Moeur, J. Shafto, D. Swanson, T. Ukrainczyk, C. Xu, D. Little . . . 75

Protein Array Technology: The Tool to Bridge Genomics and Proteomics
H. Eickhoff, Z. Konthur, A. Lueking, H. Lehrach, G. Walter, E. Nordhoff,
L. Nyarsik, K. Büssow 103

Microarray Data Representation, Annotation and Storage
A. Brazma, U. Sarkans, A. Robinson, J. Vilo, M. Vingron, J. Hoheisel,
K. Fellenberg 113

Author Index Volumes 51 – 77 141

Subject Index 153

Combinatorial Algorithms for Design of DNA Arrays

Sridhar Hannenhalli[1] · Earl Hubbell[2] · Robert Lipshutz[2] · Pavel A. Pevzner[3]

[1] Departments of Mathematics, University of Southern California, Los Angeles, California
90089-1113, USA. Present address: Celera Genomics, Rockville, Maryland 20850, USA
[2] Affymetrix, 3380 Central Expressway Santa Clara, California 95051, USA.
E-mail: earl_Hubbell@affymetrix.com
[3] Departments of Computer Science and Engineering, University of California at San Diego,
La Jolla, California 92093-0114, USA. E-mail: ppevzner@cs.ucsd.edu

Optimal design of DNA arrays requires the development of algorithms with two-fold goals:
reducing the effects caused by unintended illumination (*border length minimization problem*)
and reducing the complexity of masks (*mask decomposition problem*). We describe algorithms
that reduce the number of rectangles in mask decomposition by 20–30% as compared to a
standard array design under the assumption that the *arrangement* of oligonucleotides on the
array is fixed. This algorithm produces provably optimal solution for all studied real instances
of array design. We also address the difficult problem of finding an arrangement which minimizes the border length and come up with a new idea of *threading* that significantly reduces
the border length as compared to standard designs.

Keywords: DNA arrays, Photolitography, Mask design, Combinatorial algorithms, Rectangle
cover problem

1   Introduction   2

2   Placement Problem   3

3   DNA Arrays and the Traveling Salesman Problem   5

4   Threading   6

5   Rectangle Cover Problem   12

6   Mask Decomposition Problem   14

7   Conclusions   18

8   References   18

Advances in Biochemical Engineering/
Biotechnology, Vol. 77
Managing Editor: T. Scheper
© Springer-Verlag Berlin Heidelberg 2002

1
Introduction

DNA arrays have been greatly successful in a variety of genomic analyses [2], ranging from detecting SNPs [13] to functional genomics [3, 9, 14]. One method of synthesizing an oligonucleotide array is by a photolithographic *V LSIPS (very large scale immobilized polymer synthesis)* method similar to that used in the semiconductor industry. In this method, light is used to direct synthesis of oligonucleotides in an array. In each step, light is selectively allowed through a mask to expose cells in the array, activating the oligonucleotides in that cell for further synthesis. For every synthesis step there is a *mask* with corresponding open (allowing light) and closed (blocking light) cells. Each mask corresponds to a step of combinatorial synthesis described in [5].

The proper regions are activated by illuminating the array through a series of masks. Unfortunately, because of diffraction, internal reflection, and scattering, points close to the border between an illuminated region and a shadowed region are often subject to unintended illumination [5]. In such a region, it is uncertain whether a nucleotide will be appended or not. This uncertainty gives rise to probes with unknown sequences complicating interpretation of the experimental data. Methods are being sought to minimize the lengths of these borders so that the level of uncertainty is reduced. The problem is important not only for Affymetrix arrays but also for any other in-situ synthesis scheme, such as the recently published micromirror arrays [12].

Border Minimization Problem (BMP) is to arrange the probes on the array in such a way that the overall border length of *all* masks is minimal. For two probes x and y, let $\delta(x, y)$ be the Hamming distance between probes, i.e., the number of positions in which x and y differ. An intuitive approach to BMP is to position similar probes at neighboring sites. In a *synchronous* array design when every iteration generates a fixed nucleotide at i-th positions of the probes, the overall border length of all masks equals $\sum \delta(x, y)$, where the sum is taken over all pairs of neighboring probes on the array. For uniform arrays this observation establishes the connection between minimization of border length and Gray codes [11]. An l-bit *Gray* code is defined as a permutation of the binary numbers between 0 and $2^l - 1$ such that neighboring numbers have exactly one differing bit, as do the first and last numbers.

For a fixed set of n probes, consider an n-vertex complete graph with edge weights equal to Hamming distances between probes. For uniform arrays with 4^l probes, every solution of the *Traveling Salesman Problem (TSP)* corresponds to a Gray code and leads to arrangements that position similar probes in neighboring sites thus minimizing the border length of the mask for uniform arrays [11]. However, for *assay* arrays with arbitrary probes used in functional genomics, the border length minimization remained an open proplem.

The goal of positioning similar probes in neighboring sites led to the attempts of using TSP approach for border length minimization in the design of dedicated functional genomics arrays. In the very first design of such arrays at Affymetrix in mid 1990s Earl Hubbell implemented a greedy TSP algorithm for minimizing the border length of the assay chip. The motivation for using a TSP tour is that consecutive probes in the tour are likely to be similar. The probe arrangement corresponding to found TSP solution was further *threaded* on an array in a row-by-row fashion. The deficiency of this approach is that although probes in the same row are usually similar, the neighboring probes in the same column may be very different thus leading to an increase in border length. In this paper we propose a different idea for array design called *optimal threading*. Define a threading as a self-avoiding path through the sites of an array. Different threadings vary in the border length and we show that a clever choice of threading reduces the border length by 20 % for large chips (as compared to the row-by-row threading).

Masks for DNA arrays are built by patterning equipment which generates rectangular regions by a long series of "flashes", each flash producing a rectangle. Complex masks are made by consecutive generation of rectangles. The cost of a mask is (approximately) linear in the number of rectangles generated, and there is a fixed upper limit to the number of rectangles it is feasible to generate, before a different (expensive) mask fabrication technology must be used instead. Therefore, we are interested in a cover of a mask with a minimum number of rectangles (the rectangles in a cover can overlap).

The rectangle cover problem is known to be NP-hard and approximation algorithms with small performance ratios for this problem are unknown ([6, 7]). We explore the specifics of oligonucleotide masks and devise an efficient algorithm which finds a (provably) optimal rectangle cover for all masks we tested at Affymetrix.

The paper is organized as follows. In the following section we present the "Placement problem" which is a generalization of a few interesting optimization problem, including optimal mask design and (surprisingly) protein folding in the lattice model. We further describe applications of TSP for DNA array design and present threading as novel approach to optimizing the border length. Finally, we describe the algorithms for optimal mask decomposition into rectangles.

2
Placement Problem

Let $G_1(V_1, E_1, w_1)$ and $G_2(V_2, E_2, w_2)$ be two complete edge-weighted graphs with weight functions w_1 and w_2. A bijective function $\psi: V_2 \to V_1$ is called a *placement* of G_2 on G_1. The cost of placement is defined as

$$c(\psi) = \sum_{x, y \in V_2} w_1(x, y) \, w_2(\psi(x), \psi(y)).$$

The *optimal placement* problem is to find a placement of G_2 on G_1 of minimal cost.

For optimal mask design problem the $N \times N$ vertices of the *grid* graph G_1 correspond to a $N \times N$ array. The distance between grid cells (i_1, j_1) and (i_2, j_2) in the *Manhattan metric* is defined as $|i_2 - i_1| + |j_2 - j_1|$. The vertices of the grid are *neighbors* if the distance between them is 1. The weight $w_1(x, y)$ is defined as 1 if x and y are neighbors and 0 otherwise. The vertices of the *probe* graph G_2 correspond to $N \times N$ probes to be synthesized on a chip. The weight function $w_2(p, q)$ is de-

Table 1. Performance comparison of various algorithms for real masks

Chip Size	549a8 312×312	549a14 312×312	549a18 312×312	549a19 312×312	549a22 312×312
#Black Cells	17377	11152	14279	32519	16102
#Maximal Rectangles	1394	993	2362	2952	3425
#Prime Rectangles	1392	968	2327	2920	3387
#Black Cells after Phase 1	4	260	610	509	433
#iterations in Phase 2	1	1	1	1	1
#Rectangles found in Phase 2	2	17	35	32	34
#Rectangles found after Phase 2	0	0	0	0	0
#Rectangles found by Rectangle_Cover	1394	985	2362	2952	3421
#Rectangles found by Rectangle_Partition	1394	985	2507	3406	3521
#Rectangles found by Rectangle_Greedy	1513	1432	2980	4588	3816

fined as the *distance* between probes p and q. The distance between probes is the number of times an *edge* occurs between them in the entire set of masks to fabricate the chip. Depending on the synthesis strategy the distance is either Hamming distance or the distance defined by the synthesis-schedule. In such formulation the optimal placement attempts to position similar probes in the neighboring position of the grid and minimizes the *overall border length* of the masks. In many cases the overall border length is well-correlated with the number of flashes to produce the mask and therefore, the optimal placement reduces the cost of the mask set in these cases (Table 1).

TSP is a particular case of the optimal placement problem (in this case G_1 is a graph with edges of weight 1 forming a path and all other edges of weight 0). This observation motivates the following approach to optimal mask design problem: find TSP in the probe graph and *thread* it in a row-by-row fashion on the array grid. Such approach optimizes the placement for half of the edges (for adjacent cells in the same row) in the grid but completely ignores another half (for adjacent cells in the same column). Experiments suggest that this approach reduces the cost of the placement by 27% as compared to the random design for 64 Kb chip. The statistical analysis indicates that the larger is the chip the larger is the saving in cost (four-fold increase in the size of the chip increase the saving by about 5%). Our goal is to optimize the placement for the remaining half of the edges and to reduce the cost of the edges by additional 27%. Below we show how to reduce it by 20% thus achieving savings which is close to optimal.

3
DNA Arrays and the Traveling Salesman Problem

Since any improvement of the TSP-tour would likely improve the parameters of array design, we first tried to find a better TSP algorithm for assay arrays. Computing an optimal TSP tour for a graph with 64,000 vertices is not feasible. Note that only very fast approximation algorithms (i.e. at most quadratic) are suitable for our size of the problem. We implemented 2-opt algorithm [8] for TSP. Given an ordering of vertices $\pi = \pi_1 \ldots \pi_{i-1}, \pi_i, \pi_{i+1} \ldots \pi_{j-1}, \pi_j, \pi_{j+1} \ldots \pi_n$, a *reversal* of π at i, j is a permutation $\pi' = \pi_1 \ldots \pi_{i-1}, \pi_j, \pi_{j-1} \ldots \pi_{i+1}, \pi_i, \pi_{j+1} \ldots \pi_n$ obtained from π by reversing the fragment $\pi_i, \pi_{i+1} \ldots \pi_{j-1}, \pi_j$. *2-opt neighborhood* of π is formed by all permutations obtained from π by reversals. The Lin–Kernighan local improvement algorithm finds a suboptimal TSP tour by starting from an arbitrary permutation of vertices $\pi = \pi(0)$. At a general step, given a permutation $\pi(i)$, it searches for a permutation $\pi(i+1)$ from 2-opt neighborhood of $\pi(i)$ such that the cost of $\pi(i+1)$ is less than the cost of $\pi(i)$. The computation terminates at a local minimum when no such permutation in 2-opt neighborhood exists.

Disappointingly, we found that a 2-opt optimization of the greedy TSP-tour leads to only 0.3% improvement (Table 1). We tried a few approaches to further improve the TSP solution (or to verify its proximity to the optimal solution), in particular, solving *minimum length cycle cover* problem with further merging of cycles in the cover into a hamiltonian path. A cycle cover is a set of (simple) cycles such that each vertex in the graph is contained in exactly one cycle. The

Table 2. Performance comparison of various algorithms for random masks with varying density; the parameter p denotes that probability of a grid cell being a black cell

Chip Size	r(p=0.9) 500×500	r(p=0.75) 500×500	r(p=0.5) 500×500	r(p=0.25) 500×500	r(p=0.1) 500×500
#Black Cells	224660	186891	123951	60791	24241
#Maximal Rectangles	50944	65399	61364	41092	20362
#Prime Rectangles	8452	28608	48399	39627	20322
#Black Cells after Phase 1	97522	37771	6400	339	24241
#iterations in Phase 2	7	11	5	2	1
#Rectangles found in Phase 2	1343	9599	3513	272	1
#Rectangles found after Phase 2	12561	2393	18	0	0
#Rectangles found by Rectangle_Cover	22356	40600	51930	39899	20323
#Rectangles found by Rectangle_Partition	22726	43633	53123	39977	20323
#Rectangles found by Rectangle_Greedy	28420	45900	55077	40267	20337

weight of the cycle cover is the total length of its cycles. A minimum weight cycle cover can be reduced to *matching* problem in the bipartite graph. To solve the cycle cover problem we used Ed Rothberg's implementation of Gabow's weighted matching algorithm (available at dimacs.rutgers.edu/pub/netflow/maxflow). However, for such large matching problems this software takes a lot of time even for 9Kb arrays and becomes prohibitively slow for larger arrays.

The cost of the minimum length cycle cover provides a lower bound on the length of TSP-tour. Since cycle cover for large graphs is hard to compute we used the following lower bound that is just slightly worse than the cycle cover:

$$\frac{1}{2} \sum_v \text{ sum of costs of two shortest edges at vertex } v,$$

where the sum is taken over all vertices of the graph. For 9K chips this bound is just 0.4% worse than the bound based on the minimum length cycle cover problem. This bound demonstrates that the greedy algorithm is already within 5% of the optimal solution while the best algorithm we developed (*Global Greedy*) is within 3% of the optimal solution (Table 1). An explanation for such an excellent performance of the greedy algorithm is that the distances between probes in the assay arrays are "almost random" (compare with [1]). *Global_Greedy* (pseudo-code shown below) algorithm proved to be the best in our tests of different approaches for TSP problems associated with DNA arrays.

Input: Complete edge weighted graph $G(V, E)$
Output: Path graph $H(V, E')$
Algorithm *Global_Greedy*
1. $E' \leftarrow \Phi$
2. while $H(V, E')$ is not connected
3. $e \leftarrow$ the shortest edge in E
4.   if $H(V, E' \cup \{e\}$ has a cycle
                    or a vertex of degree 3
5.   delete edge e from E
6.   else
7. $E' \leftarrow E' \cup \{e\}$
8. Output $H(V, E')$

4
Threading

Since further improving of TSP tour is infeasible we turned to another idea. In this section we show how a choice of threading could significantly improve the border length for a given TSP tour over the set of probes. Figure 1 presents two different threadings of the same TSP tour. The neighbors on the grid that are not adjacent along the TSP tour are likely to have high cost. We refer to them as "bad" edges. At first glance it looks like threading does not affect the cost of placement. Figure 1a has nine "bad" (vertical) edges (shown by dashed and dotted lines). Figure 1b has nine other "bad" edges (shown by dashed and dotted lines). One can assume that the average cost of a "bad" edge is roughly the expected Ham-

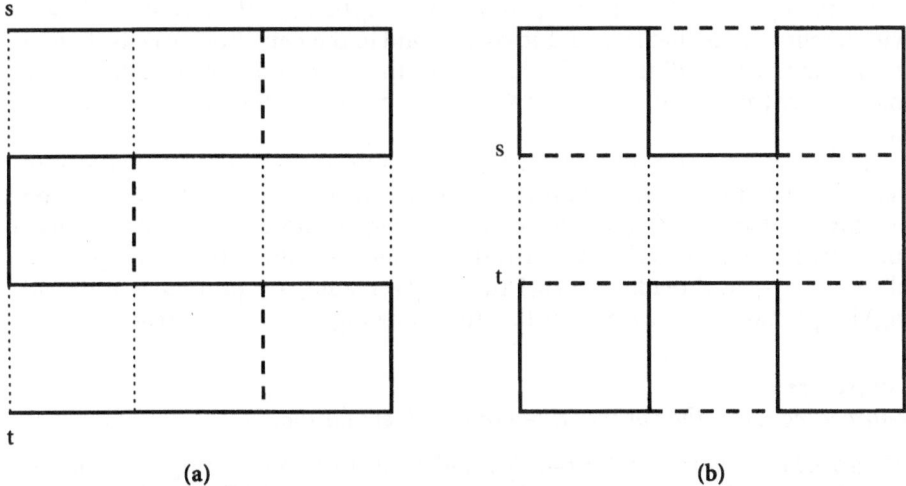

Fig. 1. Threadings (shown by *solid edges*) of the same TSP tour on 4 × 4 array: a standard row-by-row threading and **b** 1-threading

ming distance between random probes since no optimization on these edges was done. However, it is not true for dashed edges in Fig. 1 b. Although no explicit optimization on dashed edges was done, we implicitly minimized their cost while searching for a TSP tour since dashed edges join "close" vertices in the TSP tour (vertices i and $i+3$). An observation that the distance between close vertices in the TSP tour is generally lower than the average distance implies that the dashed edges correspond to relatively small distances. Therefore, some implicit optimization on these edges was done. Figure 1 a has 3 dashed edges while Fig. 1 b has 6 dashed edges thus indicating that threading in Fig. 1 b is better. This observation leads to the following problem.

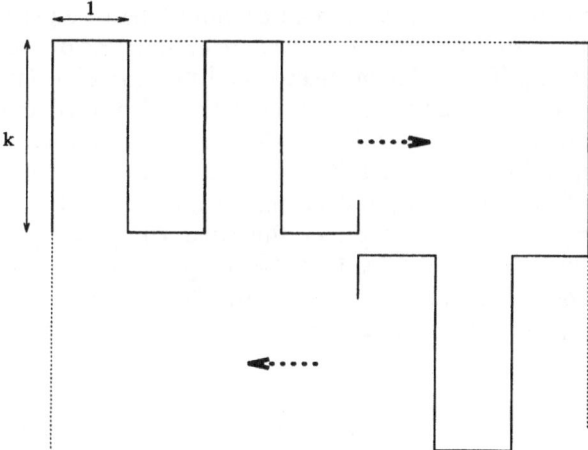

Fig. 2. k-Threading

Threading is a (self-avoiding) path $\pi = \pi_1 \ldots \pi_n$ through the elements of a grid with n cells. Notice that if π_i and π_j correspond to the neighboring cells then $j - i$ is odd. Given a threading π and a threshold t, the vertices π_i and π_j in the threading are *close* (or t-close) if $1 < |j - i| \leq t$ and the corresponding cells are neighbors in the grid.

Optimal threading problem (OTP) is to find a threading maximizing the number of close vertices. Below we present the solution of the optimal threading problem for small values of t ($t = 3, 5, 7$) that are adequate for arrays commonly used functional genomics. Define a k-threading as shown in Fig. 2. 0-threading is simply a row-by-row threading. If $t = 1$ then any threading (in particular, 0-threading) is optimal. It fact, we believe that the following conjecture is true:

Conjecture
k-threading solves the optimal threading problem for the threshold $t = 2k + 1$.

We proved this conjecture for $t = 3, 5, 7$ and came up with rather tight bounds for $k > 7$. Surprisingly, the proofs for all these three cases use different ideas.

Theorem 1
1-*Threading provides an optimal threading for OTP with the threshold 3.*

Proof
We bound the number of close vertices in a threading $\pi = \pi_1 \ldots \pi_n$. Assign to a threading π a sequence $a = a_1 \ldots a_n$ such that $a_i = 1$ if π_i is close to π_{i+3}, and $a_i = 0$ otherwise. Simple geometric analysis demonstrates that if π_i and π_{i+3} are close then π_{i+1} and π_{i+4} can not be close and π_{i-1} and π_{i+2} can not be close. Therefore, every 1 in the sequence a is preceeded and followed by a 0. Therefore, the number of 1's in a (and the number of close vertices in the threading π) is at most $n/2$. On the other hand, the number of close vertices in 1-threading approaches $n/2$ thus proving that 1-threading is optimal.   □

For a given threading $\pi = \pi_1 \ldots \pi_n$ and a threshold t define a graph on vertices $\{\pi_1, \ldots, \pi_n\}$ with π_i and π_j adjacent iff they are either close or consecutive vertices in the threading (Fig. 3). For vertices π_i and π_{i+2k+1} define (π_{i+k}, π_{i+k+1}) as the *middle edge* between π_i and π_{i+2k+1}. Every pair of close vertices in the threading graph imposes a *unit load* on the middle edge between them. For example, close vertices π_1 and π_6 impose a unit load on the edge (π_3, π_4) in the threading graph (Fig. 3). Close vertices π_2 and π_5 impose another unit load on the edge (π_3, π_4) thus imposing overall load 2 on this edge. The overall load of all close vertices of the threading in Fig. 3 is 12. Below we prove that for the threshold 5, the overall load of any threading is bounded by $2/3\, n$ thus demonstrating that 2-threading is asymptotically optimal for $t = 5$.

Theorem 2
2-*Threading provides an optimal threading for OTP with the threshold 5.*

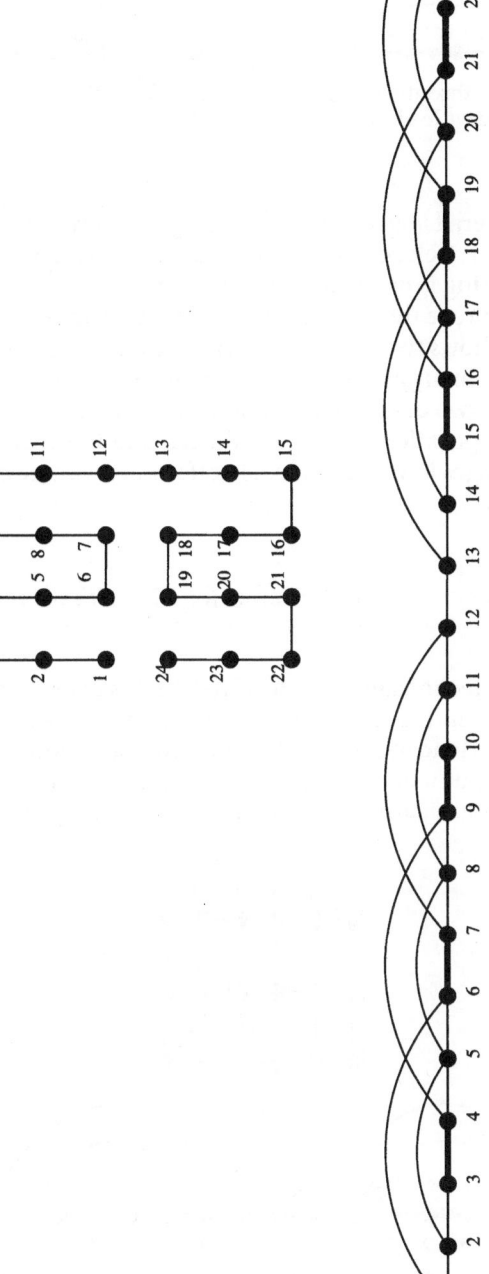

Fig. 3. Threading graph of a 2-threading with the threshold $t = 5$. Bold edges have load 2

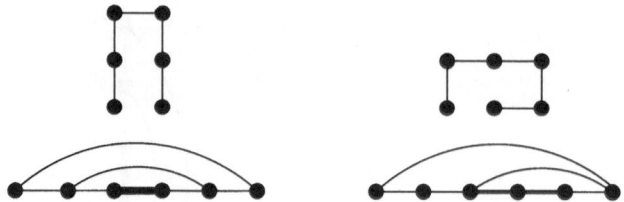

Fig. 4. Fragments of the threading graphs for two possible configurations of close vertices at distance 5. Bold edges have load 2 or 1

Proof

We prove that the overall load of every threading with threshold 5 is at most 2/3 n. Since the number of 5-close vertices in 2-threading is 2/3 n asymptotically it proves that 2-threading is optimal for this threshold.

It is sufficient to prove that the overall load of every three consecutive edge is at most 2. Figure 4 shows two different configurations for close edges at distance 5. Simple geometric analysis demonstrates that for the first configuration two edges to the left and two edges to the right from the loaded edge have zero load. For the second configuration the edge to the left and the edge to the right from the pair of loaded edges have zero load. Therefore, every three consecutive edges have load at most 2.   □

Theorem 3
3-Threading provides an optimal threading for OTP with the threshold 7.

Proof

The proof for $t=7$ is technically more involved. In this case, we have to prove that the number of edges in the threading graph is at most 3/4 n but the argument "every 4 edges have a load at most 3" does not work anymore (Fig. 5). We briefly sketch a different argument which proves the optimality of 3-threading in this case. A sequence of consecutive edges in the threading graph is a *run* if each edge

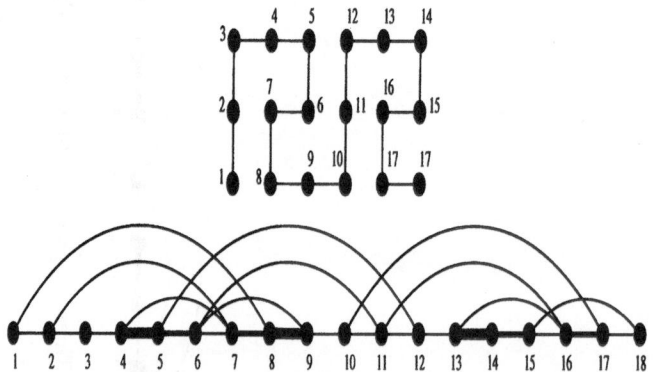

Fig. 5. An example of a threading graph with load 4 on four consecutive edges joining vertices 4 and 7 ($t=7$). Thick bold edges have load 2 while thin bold edges have load 1

in this sequence has a non-zero load and the edges to the right and to the left of this sequence have a zero load. Every run with m edges has $l > 0$ zero-load edges to the left and $r > 0$ zero-load edges to the right. For each run define the parameter $p = m + \dfrac{l+r}{2}$. It is easy to see that if for every run the parameter

$$\frac{load\ of\ the\ run}{p} \le \frac{3}{4}$$

then the number of close edges in the threading graph is smaller then $3/4\,n$. For example, for a run of edges $(4, 5)$ and $(5, 6)$ with loads 2 and 1 correspondingly, $l = 3, m = 2, r = 1$. Therefore

$$\frac{load\ of\ the\ run}{p} = \frac{2+1}{2+\dfrac{3+1}{2}} \le \frac{3}{4}.$$

A case-by-case analysis of different configurations with $t = 7$ demonstrates that

$$\frac{load\ of\ the\ run}{p} \le \frac{3}{4}$$

for every run (proof is omitted).   □

Table 3. Performance of 1-threading

	DataSet 1	DataSet 2	DataSet 3	DataSet 4	DataSet 5
1. Size	4K	9K	14K	16K	66K
2. Chip Dimensions	134*134	134*134	256*256	256*256	262*262
3. A lower bound on a TSP tour cost	28246	57836	96513	60815	221995
4. Local-greedy TSP tour cost	30045	61229	100675	64392	236835
5. 2-opt improved TSP tour cost	30012	61187	100657	64213	235985
6. Global-greedy TSP tour cost	29367	60124	99108	63679	234356
7. Maximum % improvement in TSP tour cost	5.9	5.5	4.1	5.3	5.9
8. Achieved % improvement in TSP tour cost	2.1	1.7	1.5	0.8	0.7
9. Edge-Cost of 0-threading	92483	195959	308021	300392	1226529
10. Edge-Cost of 1-threading	86764	181639	293652	253420	1005429
11. Edge-Cost of improved 1-threading	86185	180653	291079	251761	993502
12. Percentage decrease in Edge-cost	6.8	7.8	5.5	16.1	20.0
13. Flash-Cost of 0-threading	29639	60481	99207	62054	228823
14. Flash-Cost of 1-threading	35356	72939	118237	57190	208059
15. Percentage change in Flash-cost	+19.2	+20.6	+19.2	−7.8	−9.1

For the experimental data considered, $t=3$ seems the most appropriate, leading to 1-threading. Notice that this threading does not optimize the costs of vertical edges between the rows $2i$ and $2i+1$. To further improve this threading we apply the Lin-Kernighan [8] optimization on the pairs of rows of the grid (every vertex in the corresponding TSP is a pair of consecutive rows $2i+1$ and $2i$). The improvement on this latest step is rather small (Table 1) thus indicating that 1-threading already provides a good solution.

5
Rectangle Cover Problem

After the arrangement of oligonucleotides on the array is fixed, the resulting masks need to be generated by patterning equipment as unions of small rectangles (rectangle cover problem). In this section we describe the *Rectangle_Cover* algorithm that takes into account the specifics of oligonucleotide masks and leads to a 10%–30% improvement in the cover size as compared to a greedy strategy.

We define a $n \times n$ grid as a collection of n^2 cells and $2n(n+1)$ edges. A *mask* is defined as a subset of the cells of a grid. Cells belonging to the mask are called *black* cells while the remaining cells are called *white* cells (Fig. 8). A *rectangle* is a set of black cells forming a rectangular shape. Given a mask, the rectangle cover problem is to find a minimum number of rectangles covering all black cells of the mask.

An edge in a grid is a *border* edge for a mask if it separates a black cell from a white cell. A rectangle in a mask is *maximal* if each of its four sides contains a border edge. A black cell is called a *leaf* if it belongs to only one maximal rectangle in a mask. A maximal rectangle is prime if it contains a leaf.

Two horizontal (vertical) edges are called *co-horizontal* (*co-vertical*) if they occupy the same column (row) of the grid. Each black cell x defines a maximal horizontal continuous strip of black cells containing x. This strip is bounded by co-vertical border edges $left(x)$ on the left and $right(x)$ on the right. The co-horizontal border edges $top(x)$ and $bottom(x)$ are defined similarly. For a cell x in a mask M a rectangle defined by border edges $left(x)$, $right(x)$, $top(x)$ and $bottom(x)$ is called the *extended* rectangle for x and is denoted $R(x, M)$. Note that an extended rectangle might contain both black and white cells. The following lemma characterizes leaves and prime rectangles.

Lemma 1
A cell $x \in M$ is a leaf iff $R(x, M)$ contains only black cells. A rectangle in a mask M is prime iff it has a pair of co-horizontal and a pair of co-vertical border edges in M.

Clearly, for every mask there exists an optimal cover which contains only maximal rectangles. It implies that there exists an optimal cover containing all prime rectangles. This observation motivates the first stage of the algorithm *Rectangle_Cover* (see [6]) for covering mask M (lines 1–2 in Fig. 6). The second stage (lines 3–11) is motivated by the following lemma.

      Algorithm *Rectangle_Cover(M)*

1.   Generate the set $\mathcal{R}$ of all prime rectangles in M
2.   "Trim" mask M by deleting all its prime rectangles:
   $$M' \leftarrow M \setminus \cup_{R \in \mathcal{R}} R$$
3.   *switch* $\leftarrow 1$
4.   **while** (*switch* $= 1$)
5.   *switch* $\leftarrow 0$
6.   **for** every black cell $x \in M'$
7.   construct extended rectangle $R(x, M)$
8.   **if** there exists a rectangle
   $\overline{R}$ such that $M' \cap R(x, M) \subset \overline{R} \subset M$
9. $\mathcal{R} \leftarrow \mathcal{R} \cup \{\overline{R}\}$
10. $M' \leftarrow M' \setminus \overline{R}$
11.   *switch* $\leftarrow 1$
12.   Cover M' by maximal rectangles in M
    using the greedy *Set_Cover* algorithm.
13.   Output all rectangles from $\mathcal{R}$
    and all rectangles found by *Set_Cover*.

Fig. 6. Algorithm *Rectangle_Cover*

Lemma 2

Let $\mathfrak{R}$ be an optimal cover of mask M and let $\mathfrak{R}' \subset \mathfrak{R}$. Define $M' = M \setminus \cup_{R \in \mathfrak{R}'} R$. Let $R(x, M)$ be the extended rectangle of cell $x \in M'$. If there exists a rectangle $\bar{R}$ such that $M' \cap R(x, M) \subset \bar{R} \subset M$, then there exists an optimal cover of M containing all rectangles from $\mathfrak{R}' \cup \{\bar{R}\}$.

Lemma 2 ensures that a set of rectangles $\mathfrak{R}$ constructed by the algorithm *Rectangle_Cover* is contained in an optimal cover. For all oligonucleotide masks constructed at Affymetrix, $\mathfrak{R}$ is an optimal cover. To complete the cover construction for the cases when $\mathcal{M}'$ is non-empty, we use a standard greedy *Set_Cover* [4] algorithm (although it was never invoked for real mask design).

It is typical of oligonucleotide masks that prime rectangles either fully cover the mask (thus forming the optimal cover) or leave a small number of cells uncovered. In the latter case the uncovered cells usually form a very simple mask (frequently a collection of non-overlapping rectangles) and an optimal cover for this mask can be found using lemma 2 (second stage of the algorithm). Since the optimal cover for typical oligonucleotide masks is frequently found after the first stage of *Rectangle_Cover* we describe an efficient algorithm for generating maximal and prime rectangles in detail.

Given the sets of intervals $\mathfrak{I}$ and $\mathfrak{J}$ define

$$\mathfrak{I} \cap \mathfrak{J} = \{\mathfrak{I} \cap \mathfrak{J} : \mathfrak{I} \in \mathfrak{I} \text{ and } \mathfrak{J} \in \mathfrak{J}\}.$$

Define

$$\mathfrak{I} \oplus \mathfrak{J} = ((\mathfrak{I} \cap \mathfrak{J}) \cup \mathfrak{J}).$$

The dynamic programming algorithm *Rectangle_List* (Fig. 7) generates the set of all maximal rectangles for a $n \times n$ mask M in $O(n^2)$ time. At the k-th iteration the *Rectangle_List* forms the set $\mathfrak{R}$ of all maximal rectangles positioned above the k-th row of M. $M(k)$ in *Rectangle_List* is defined as the set of intervals

Algorithm *Rectangle_List(M)*
1. $\mathcal{R} \rightarrow \emptyset$
2. $\mathcal{I} \rightarrow \emptyset$
3.   for $k \leftarrow 1$ to $n + 1$
4.   for each interval $I \in \mathcal{I}$
5.   if $I \notin \mathcal{I} \oplus M(k)$
6.   find the maximal rectangle R
                    with base I in the $(k-1)$-th row
7. $\mathcal{R} \leftarrow \mathcal{R} \cup \{R\}$
8. $\mathcal{I} \leftarrow \mathcal{I} \oplus M(k)$
9.   Output $\mathcal{R}$

Fig. 7. Algorithm *Rectangle_List*

corresponding to the k-th row of M, every such interval being a maximal continuous strip of black cells in the k-th row of M. A *base* of a rectangle is its lower side.

It is easy to see that the bases of maximal rectangles cannot partially overlap thus implying that the number of maximal rectangles is $O(n^2)$. Therefore, *Rectangle_List* can be implemented in $O(n^2)$ time. Lemma 1 implies that an algorithm to generate the set of all prime rectangles can be implemented in $O(n^2)$ time thus leading to an $O(n^2)$ implementation of the first stage of *Rectangle_Cover*. Each iteration of the second stage of *Rectangle_Cover* can be implemented in $O(n^3)$. Since the number of such iterations is $O(n^2)$, the worst-case running time of the algorithm is dominated by the second stage and is $O(n^5)$. However, for oligonucleotide masks the number of the iterations of the second stage is very small (if any).

6
Mask Decomposition Problem

Earlier attempts to minimize the flashcount generated *partitions* of the mask into rectangles (a set of *non-overlapping* rectangles covering the mask). *Rectangle_Greedy* which was used for mask design at earlier stages adds the largest possible rectangle to the already found partition while scanning the mask in left-to-right top-to-bottom fashion. Due to the regular structure of standard mask designs, this algorithm proved to be very successful.

An attempt to improve this algorithm resulted in the *Rectangle_Partition* algorithm that computes the *partition complexity* of a mask (i.e., the minimum number of non-overlapping rectangles covering a mask). *Rectangle_Partition* is based on a reduction of the rectangle partition problem to the maximum matching problem which was first devised for data compression in databases of two-dimensional pictures [10]. *Rectangle_Partition* is based on a duality theorem which is given below.

First we need a few definitions. Every mask consists of a *connected areas* and *w internal windows*. For example a mask in figure 1 consists of $a = 3$ connected components and $w = 5$ internal windows. Two points in a mask are *outerconnected*

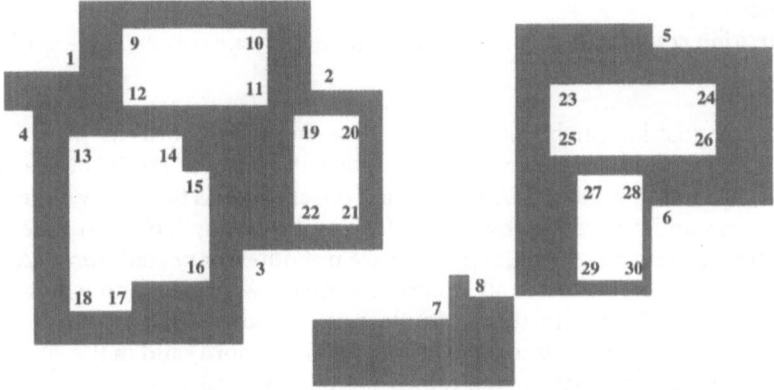

#connected components a=3, #internal windows w=5, #obtuse corners k=30
COMPLEXITY OF THE MASK c=k+a-w=30+3-5=28

OPTIMAL DECOMPOSITION OF THE MASK INTO 28 RECTANGLES:

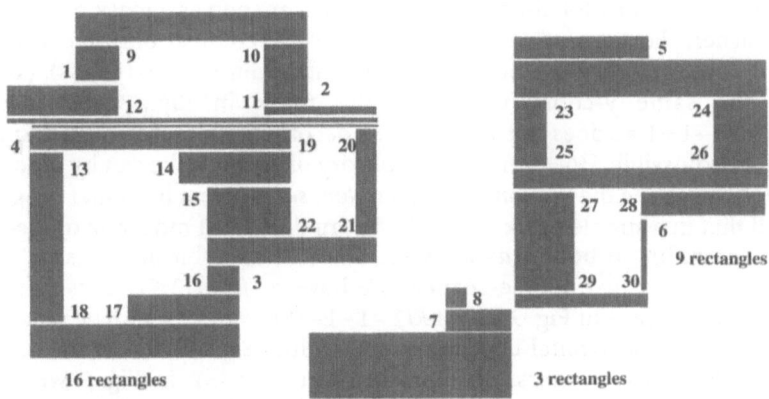

Fig. 8. Decomposition of a simple mask into rectangles

if there is a path between them that lies entirely outside the mask. For example, in Fig. 8 points 1 and 2 are outerconnected while points 1 and 12 are not. Every mask has two kinds of corners: *sharp* 90° degree corners and *obtuse* 270° degree corners. The mask shown in Fig. 8 has 22 sharp and 30 obtuse corners (obtuse corners are numbered in Fig. 8). A *chord* is a horizontal or vertical line joining two vertices on the (external or internal) border of the mask and having no other vertex on the border. Two obtuse corners are *colinear* if they have the same x or y coordinates and lie on the same chord. A mask is *simple* if no two obtuse corners are colinear. Figure 8 presents an example of a simple mask. The following theorem demonstrates that every obtuse corner and every connected component in the mask presents a potential obstacle while every window presents a potential bonus for decomposition of a simple mask.

Theorem 4
The partition complexity of a simple mask with k obtuse corners is $a - w + k$.

Proof
Mask decomposition can be interpreted as drawing some chords in the mask (see Fig. 8). After a chord is drawn we assume that it creates a "strait" between two parts of the mask it separates. An important observation is that every chord joining vertices x and y increases the parameter $a - w$ by exactly 1. If x and y are outer-connected than $\Delta a = 1$, $\Delta w = 0$, if x and y are not outerconnected $\Delta a = 0$, $\Delta w = -1$ resulting in $\Delta(a - w) = 1$. After all m chords were drawn there is no windows left and $a - w + m = c$. Every chord in a simple mask can destroy at most one obtuse corner (since no two obtuse corners lay on the same chord) and in the end all obtuse corners should be destroyed (since there is no obtuse corner in a decomposition of the mask into rectangles). It implies that $k \leq m$ and $a - w + k \leq c$. One can see that making k horizontal chords (one for each obtuse corner) destroys all obtuse corners and decomposes the simple mask into rectangles (Fig. 8). Therefore for a simple mask $c = k + a - w$.   □

Theorem 4 works only for simple masks and fails as soon as a mask has colinear obtuse corners. For example, the mask in Fig. 9 has colinear obtuse corners (2, 13), (3, 9) and (4, 8) having the same x-coordinates and (1, 5), (15, 11), (14, 12) having the same y-coordinates. For the mask in Fig. 9 the formula $k + a - w = 19 - 1 + 1 = c$ does not work since a decomposition of this mask into 15 rectangles is possible. However, the complexity of such masks can be efficiently computed by solving the *maximum independent set* problem in a bipartite graph.

Recall that in a simple mask every chord can destroy at most one obtuse corner thus providing a bound $a - w + k \leq c$. Since in an arbitrary mask a chord can destroy up to two obtuse corners we have $a - w + k/2 \leq c$. This bound is over-optimistic, since in Fig. 9 $a - w + k/2 = 1 - 1 - 19/2 = 10 < c = 15$. The reason for this is that some horizontal chords compete with vertical chords for the best utilization of obtuse corners. This competition is shown in Fig. 9 where the horizontal chord (1,5) competes with the vertical chord (2,13). We introduce the *chord graph* $G(V, E)$ of the mask where V is the set of all chords joining two obtuse corners and $E = \{(v, w): \text{chords } v \text{ and } w \text{ do not cross}\}$. Every maximum *independent set* in this graph provides a strategy for destroying obtuse corners. Denoting the size of a maximum independent set in G as i, the minimal number of chords which needs to be drawn to decompose a mask into rectangles is $i + (k - 2i) = k - i$ (since each chord from the independent set destroys two obtuse corners and after these chords are drawn there are $k - 2i$ obtuse corners left). Using Theorem 4 one can show that every obtuse corner and every connected component presents a potential obstacle while every window and every vertex in the maximal independent set presents a potential bonus for decomposition of an arbitrary mask:

Theorem 5
The partition complexity of an arbitrary mask is $c = a - w + k - i$.

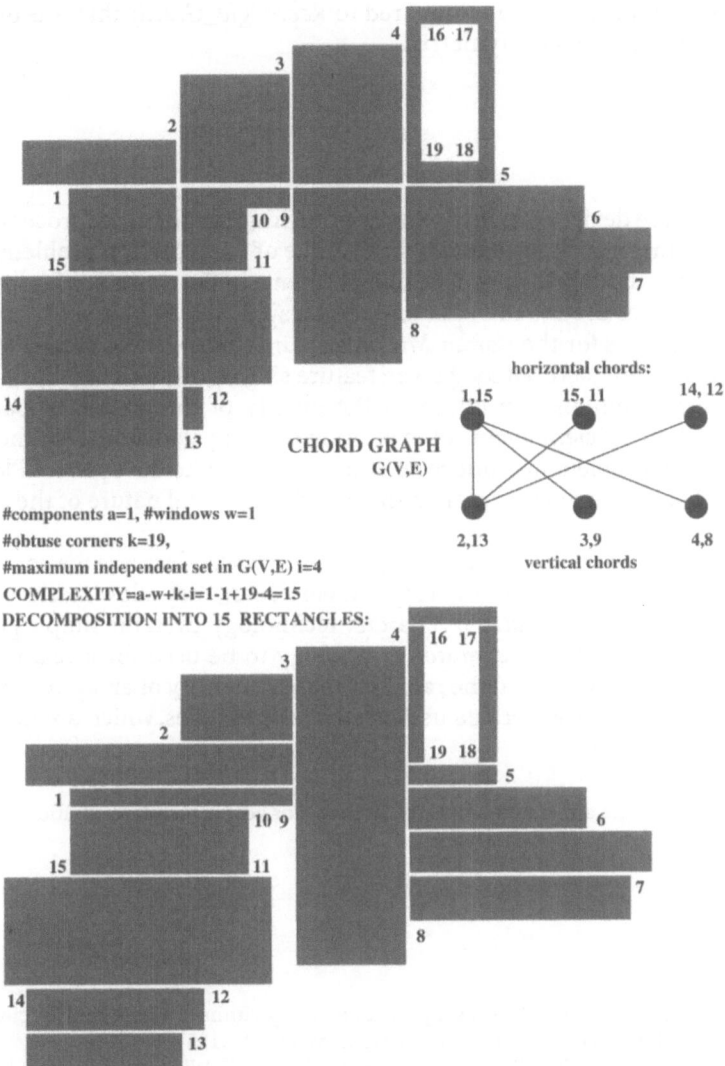

Fig. 9. Chord graph and decomposition of mask into rectangles

Theorem 5 is the basis of the *Rectangle_Partition* algorithm that reduces mask decomposition to two classical graph-theoretical problems: computing the number of connected components in a graph (to compute *a* and *w*) and computing the size of maximal independent set in the bipartite graph.

Rectangle_Cover was implemented and compared with *Rectangle_Greedy* and *Rectangle_Partition*. Recall that *Rectangle_Greedy* and *Rectangle_Partition* generate a *partition* of a mask into rectangles (a set of *non-overlapping* rectangles covering the mask). The tables 6 and 3 compares the algorithms for both real masks and randomly generated masks. The Rectangle_Cover algorithm produces up to a 30% improvement as compared to *Rectangle_Greedy* that was originally used for array design at Affymetrix.

7
Conclusions

DNA arrays are devices of complexity approaching that of microprocessors. The manufacturing of such devices is a rich source of optimization problems. These problems lead directly to practical improvements in the process, without changing the machines used to manufacture the arrays.

The algorithms for the *Border Minimization Problem* allow probes of higher quality to be manufactured for a given feature size, by minimizing the amount of light leakage. This directly improves the quality of the arrays. In particular, *threading* is an efficient means of exploiting existing algorithms for the Travelling Salesman Problem to produce an efficient solution to the *Optimal Placement Problem*. The recent work exploited the two-dimensional nature of the array directly.

The algorithms for the *Mask Decomposition Problem* decrease the cost of mask fabrication, and thus the cost of arrays. In particular, they allow masks of higher density to be made with *pattern generator* technology. Even the simple *greedy* algorithm allowed *pattern generator* technology to be used for a relatively long span of time. However, given the rapid increase in density of arrays, current generations of masks are generated using *beam* technologies, which are not limited by mask complexity.

Acknowledgement. We are grateful to David Johnson for advice on approximation algorithms and software for large TSP instances.

8
References

1. Angluin D, Valiant LG (1979) Fast probabilistic algorithms for hamiltonian circuits and matchings. Journal of Computer Science and Systems 18:155
2. Chee M, Yang R, Hubbell E , Berno A, Huang XC, Stern D, Winkler J, Lockhart DJ, Morris MS, Fodor SA (1996) Accessing genetic information with high-density dna arrays. Science 274:610
3. Cho RJ, Campbell MJ, Winzeler EA, Steinmetz L, Conway A, Wodicka L, Wolfsberg TG, Gabrielian AE, Landsman D, Lockhart DJ, Davis RW (1998) A genome-wide transcriptional analysis of the mitotic cell cycle. Molecular Cell 2:65
4. Cormen TH, Leiserson CE, Rivest RL (1990) Introduction to Algorithms. MIT Press
5. Fodor S, Read JL, Pirrung MC, Stryer L, Tsai Lu A, Solas D (1991) Light-directed, spatially addressable parallel chemical synthesis. Science 251:767
6. Franzblau DS, Kleitman DJ (1984) An algorithm for covering polygons with rectangles. Information and Control 63:164
7. Garey MS, Johnson DS (1979) Computers and Intractibility: A Guide to the Theory of NP-Completeness. W. H. Freeman, San Francisco, CA
8. Lin S, Kernighan BW (1973) An effective heuristic for the traveling-salesman problem. Operation Research 21:498

9. Lockhart DJ, Dong H, Byrne MC, Follettie MT, Gallo MV, Chee MS, Mittmann M, Wang C, Kobayashi M, Horton H, Brown EL (1996) Expression monitoring by hybridization to high-density oligonucleotide arrays. Nature Biotechnology 14:1675

10. Ohtsuki, T (1982) Minimum dissection of rectilinear regions. Proc. 1982 IEEE Symp. on Circuits and Systems 1210

11. Pevzner PA (2000) Computational Molecular Biology: An Algorithmic Approach. The MIT Press instead

12. Singh-Gasson S, Green RD, Yue Y, Nelson C, Blattner F, Sussman MR, Cerrina F (1999) Maskless fabrication of light-directed oligonucleotide microarrays using a digital micromirror array. Nature Biotechnology 17:974

13. Wang DG, Fan J, Siao C, Berno A, Young P, Sapolsky R, Ghandour G, Perkins N, Winchester E, Spencer J, Kruglyak L, Stein L, Hsie L, Topaloglou T, Hubbell E, Robinson E, Mittmann M, Morris M, Shen N, Kilburn D, Rioux J, Nusbaum C, Rozen S, Hudson T, Lipshutz R, Chee M, Lander E (1998) Large-scale identification, mapping, and genotyping of single-nucleotide polymorphisms in the human genome. Science 280:1077

14. Wodicka L, Dong H, Mittmann M, Ho MH, Lockhart DJ (1997) Genome-wide expression monitoring in saccharomyces cerevisiae. Nature Biotechnology 15:1539

Received: August 2001

High-Density GeneChip Oligonucleotide Probe Arrays

Glenn H. McGall[1] · Fred C. Christians[2]

[1] Affymetrix, Inc., 3380 Central Expressway, Santa Clara, CA 95051, USA.
 E-mail: *glenn_mcgall@affymetrix.com*
[2] Affymetrix, Inc., 3380 Central Expressway, Santa Clara, CA 95051, USA.
 E-mail: *fred_christians@affymetrix.com*

High-density DNA probe arrays provide a highly parallel approach to nucleic acid sequence analysis that is transforming gene-based biomedical research. Photolithographic DNA synthesis has enabled the large-scale production of GeneChip probe arrays containing hundreds of thousands of oligonucleotide sequences on a glass "chip" about 1.5 cm² in size. The manufacturing process integrates solid-phase photochemical oligonucleotide synthesis with lithographic techniques similar to those used in the microelectronics industry. Due to their very high information content, GeneChip probe arrays are finding widespread use in the hybridization-based detection and analysis of mutations and polymorphisms ("genotyping"), and in a wide range of gene expression studies.

Keywords: GeneChip array, Oligonucleotide probe array, Photolithography, Gene expression monitoring, Genotyping

1	**Introduction**	22
2	**Array Production Technology**	24
2.1	Substrate Preparation and General Approach	24
2.2	Photolithography	26
2.3	Light-Directed Synthesis Chemistry	26
2.4	Future Enhancements	29
3	**Applications**	31
3.1	Gene Expression Monitoring	31
3.2	Genotypic Analysis	34
3.3	Other Applications	38
4	**References**	41

Abbreviations

CE	2-cyanoethyl
DMT	4,4′ dimethoxytriphenylmethyl
DLP	digital light processor

Advances in Biochemical Engineering/
Biotechnology, Vol. 77
Managing Editor: T. Scheper
© Springer-Verlag Berlin Heidelberg 2002

DTT   dithiothreitol
HPLC   high performance liquid chromatography
kb   kilobase
MeNPOC   α-methyl-6-nitropiperonyloxycarbonyl
NNEOC   1-(8-nitronaphth-1-yl)ethyloxycarbonyl
NPPOC   2-(2-nitrophenyl)propyloxycarbonyl
NTP   nucleoside 5'-triphosphate
PAG   photo-acid generator
PEO   poly(ethylene oxide)
PIV   pivaloate
pTs   para-toluenesulfonate
PYMOC   1-pyrenylmethyloxycarbonyl
SNP   single-nucleotide polymorphism
TCEP   tris(2-carboxyethyl)phosphine

1
Introduction

High-density polynucleotide probe arrays are among the most powerful and versatile tools for accessing the rapidly growing body of sequence information that is being generated by numerous public and private sequencing efforts. Consequently, this technology is having a major impact on biological and biomedical research [1, 2]. These arrays are essentially large sets of nucleic acid probe sequences immobilized in defined, addressable locations on the surface of a substrate capable of accessing large amounts of genetic information from biological samples in a single hybridization assay. In a typical application [2], DNA or RNA "target" sequences of interest are isolated from a biological sample using standard molecular biology protocols. The sequences are fragmented and labeled with fluorescent molecules for detection, and the mixture of labeled sequences is applied to the array, under controlled conditions, for hybridization with the surface probes. The array is then imaged with a fluorescence-based reader to locate and quantify the binding of target sequences from the sample to complementary sequences on the array, and software reconstructs the sequence data and presents it in a format determined by the application. Thus, in addition to the arrays themselves, the Affymetrix GeneChip system provides a fluidics station for performing reproducible, automated hybridization and wash functions; a high-resolution scanner for reading the fluorescent hybridization image on the arrays; and software for processing and querying the data (Fig. 1).

GeneChip technology is distinguishable from other microarray methods in that oligonucleotide probe sequences are photolithographically synthesized, in a parallel fashion, directly on a glass substrate. In a minimum number of synthetic steps, arrays containing hundreds of thousands of different probe sequences, typically 20 or 25 bases in length, can be generated at densities in the order of 10^5–10^6 sequences/cm^2 (Fig. 2). This capability, deliverable on a commercial production scale, is well beyond that of any other technology currently

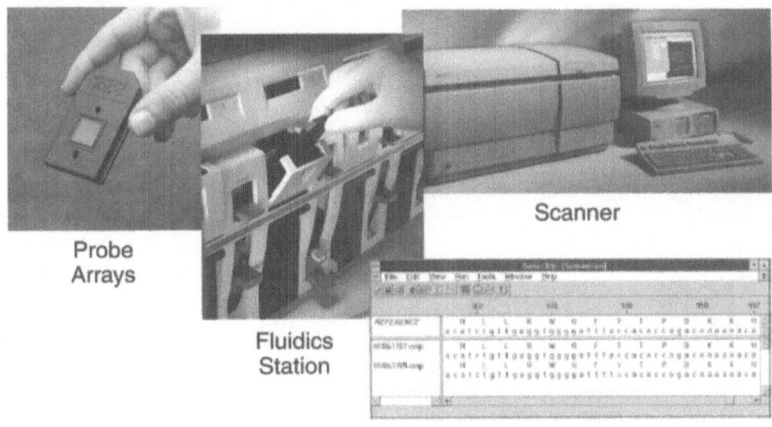

Fig. 1. GeneChip system overview

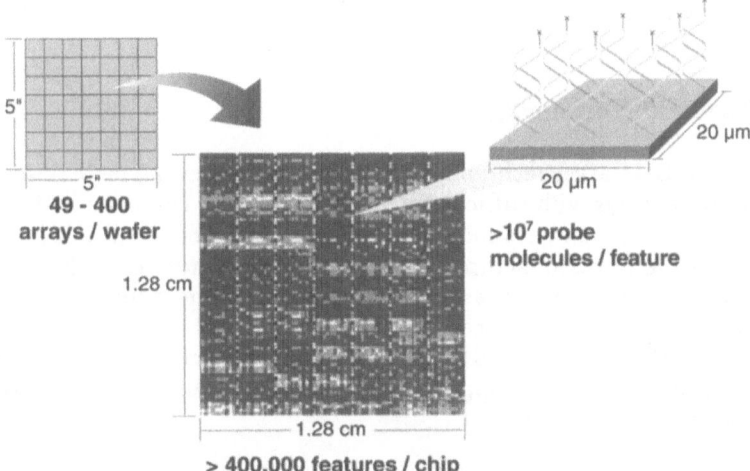

Fig. 2. Wafer-scale GeneChip production specifications

available and allows unprecedented amounts of sequence information to be encoded in the arrays.

Other array construction technologies, such as micropipetting or inkjet printing, rely on mechanical devices to deliver minute quantities of reagents to predefined regions of a substrate in a sequential fashion. In contrast, the photolithographic synthesis process is highly parallel in nature, making it intrinsically robust and scalable. This provides significant flexibility and cost advantages in terms of materials management, manufacturing throughput, and quality control. To researchers the benefits are a high degree of reliability, uniformity of array performance, and an affordable price.

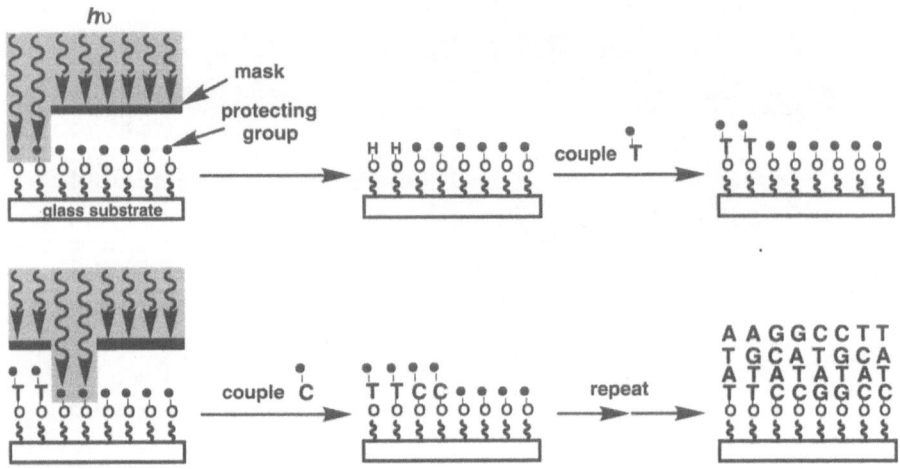

Fig. 3. Photolithographic synthesis of oligonucleotide arrays

2
Array Production Technology

The advent of DNA array technology has relied on the development of methods for fabricating arrays with sufficiently high information content and density in a rapid, reproducible and economic fashion. Light-directed synthesis [3–7] has made it possible to manufacture arrays containing hundreds of thousands of oligonucleotide probe sequences on glass "chips" little more than 1 cm² in size, and to do so on a commercial scale. In this process, 5′- or 3′-terminal protecting groups are selectively removed from growing oligonucleotide chains in predefined regions of a glass support, by controlled exposure to light through photolithographic masks (Fig. 3).

2.1
Substrate Preparation and General Approach

Prior to photolithographic synthesis, planar glass substrates are covalently modified with a silane reagent to provide a uniform layer of covalently bonded hydroxyalkyl groups on which oligonucleotide synthesis can be initiated (Fig. 4). In a second step, a photo-imagable layer is added by extending these synthesis sites with a poly(ethylene oxide) linker which has a terminal photolabile hydroxyl protecting group. When specific regions of the surface are exposed to light, synthesis sites within these regions are selectively deprotected, and thereby "activated" for the addition of nucleoside phosphoramidite building blocks.

These nucleotide precursors, also protected at the 5′ or 3′ position with a photolabile protecting group, are applied to the substrate, where they react with the surface hydroxyl groups in the pre-irradiated regions. The monomer coupling step is carried out in the presence of a suitable activator, such as tetrazole or di-

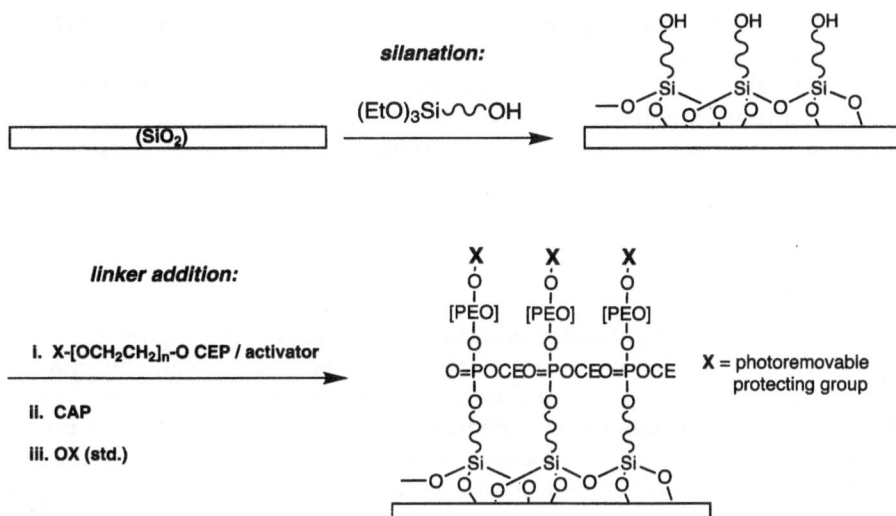

Fig. 4. Chemical preparation of glass substrates for light-directed synthesis of oligonucleotide arrays

cyanoimidazole. The coupling reaction is followed by conventional capping and oxidation steps, which also use standard reagents and protocols for oligonucleotide synthesis [5,7]. Alternating cycles of photolithographic deprotection and nucleotide addition are repeated to build the desired two-dimensional array of sequences, as described in Fig. 3.

Semiautomated cleanroom manufacturing techniques, similar to those used in the microelectronics industry, have been adapted for the large-scale commercial production of GeneChip arrays in a multi-chip wafer format (Fig. 5). Each wafer contains 49–400 replicate arrays, depending on the size of the array,

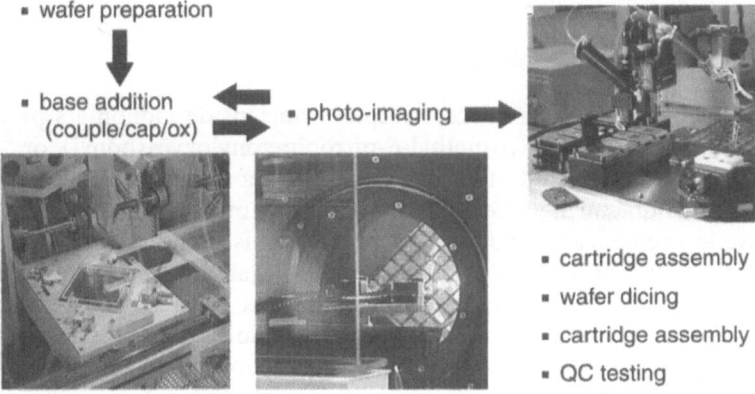

Fig. 5. Automated array manufacturing

and multiple-wafer lots can be processed together in a procedure that takes less than 24 h to complete. Multiple lots are processed simultaneously on independent production synthesizers operating continuously. After a final chemical deprotection, finished wafers are diced into individual arrays, which are finally mounted in injection-molded plastic cartridges for single-use application (see Fig. 1).

2.2
Photolithography

The photolithographic process provides a very efficient route to high-density arrays by allowing parallel synthesis of large sets of probe sequences. The number of required synthetic steps to fabricate an array is dependent only on the length of the probes, not the number of probes. A complete set, or any subset, of probe sequences of length "n" requires $4 \times n$ synthetic steps, at most. Masks can be designed to make arrays of oligonucleotide probe sequences for a variety of applications. Most arrays are comprised of custom-designed sets of probes 20–25 bases in length, and optimized masking strategies allow such arrays to be completed in as few as $3n$ steps.

The spatial resolution of the photolithographic process determines the maximum achievable density of the array and therefore the amount of sequence information that can be encoded on a chip of a given physical dimension. A contact lithography process (Fig. 3) is used to fabricate GeneChip arrays with individual probe features that are 18×18 microns in size. Between 49 and 400 identical arrays are produced simultaneously on $5'' \times 5''$ wafers. For the largest-format chip currently in production (1.6 cm^2), this provides wafers of 49 individual arrays containing more than 400,000 different probe sequences each. For arrays containing fewer probe sequences, this feature size enables more replicate arrays to be fabricated on each wafer. The technology has proven capability for fabricating arrays with densities greater than 10^6 sequences/cm^2, corresponding to features less than 10×10 microns in size. This level of miniaturization is beyond the current reach of other technologies for array fabrication.

2.3
Light-Directed Synthesis Chemistry

The current manufacturing process employs nucleoside monomers protected with a photoremovable 5'-(α-methyl-6-nitropiperonyloxycarbonyl), or "MeN-POC" group [4, 5], depicted in Fig. 6, which offers a number of advantages for large-scale manufacturing. These phosphoramidite monomers are relatively inexpensive to prepare, and photolytic deprotection is induced by irradiation at near-UV wavelengths ($\phi \sim 0.05$; $\lambda_{max} \sim 350$ nm) so that photochemical modification of the oligonucleotides, which absorb energy at lower wavelengths, can be avoided. The photolysis reaction involves an intramolecular redox reaction and does not require any special solvents, catalysts or coreactants. Since the photolysis can be performed "dry", high-contrast contact lithography can be used to achieve very high-resolution imaging. Complete photodeprotection requires less

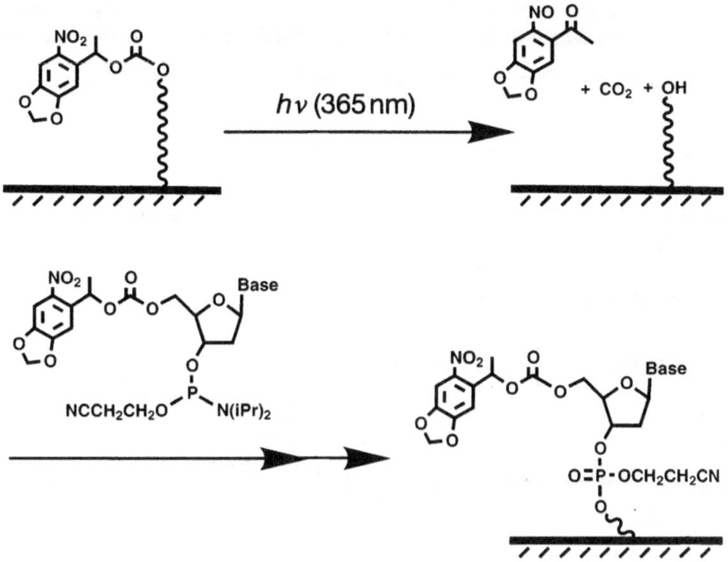

Fig. 6. Light-directed oligonucleotide synthesis cycle using MeNPOC photolabile phosphoramidite building blocks

than 1 min using filtered I-line (365 nm) emission from a commercial collimated mercury light source.

Photochemical deprotection rates and yields for oligonucleotide synthesis can both be monitored directly on planar supports using procedures based on surface fluorescence [5]. We have also developed a sensitive assay in which test sequences are synthesized on a support designed to allow the cleavage and direct quantitative analysis of labeled oligonucleotide products using ion-exchange high performance liquid chromatography (HPLC) with fluorescence detection [8, 9]. This method involves photolithographic synthesis of test sequences after the addition of a base-stable disulfide linker and a fluorescein monomer to the support (Fig. 7). The disulfide linker remains intact throughout the synthesis and deprotection, but can be subsequently cleaved under reducing conditions to release the synthesis products, all of which are uniformly labeled with a 3'-fluorescein tag. The labeled oligonucleotide synthesis products are then analyzed using HPLC or capillary electrophoresis with fluorescence detection, enabling direct quantitative analysis of synthesis efficiency. The sensitivity of fluorescence is a key feature of this methodology, since the quantities of DNA synthesis products on flat substrates are very low ($1-100$ pmol/cm^2 [9]), and difficult to analyze by other means.

The average stepwise efficiency of the light-directed oligonucleotide synthesis process is limited by the yield of the photochemical deprotection step which, in the case of MeNPOC nucleotides, is 90%–94% [5]. The other chemical reactions involved in the base addition cycles (coupling, capping, oxidation) use reagents in a vast excess over surface synthesis sites, and, provided that sufficient reagent concentrations and time are allowed for completion, they are es-

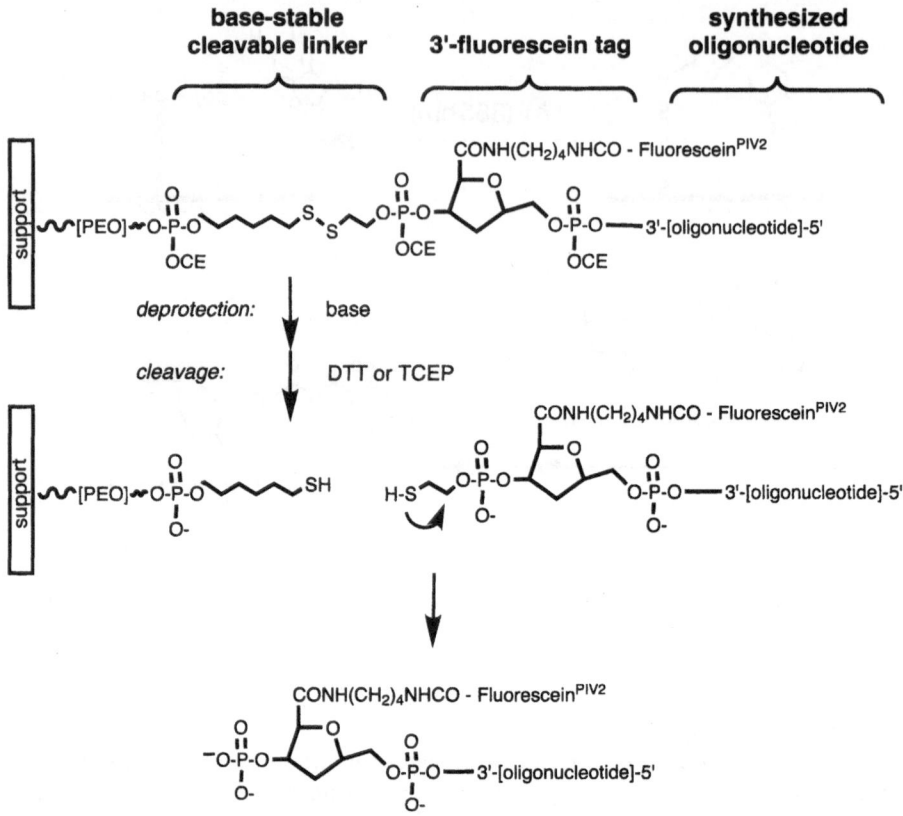

Fig. 7. Method for fluorescent labeling and cleavage of photolithographically synthesized oligonucleotides allows quantitative analysis by HPLC

sentially quantitative. However, the sub-quantitative photolysis yields lead to incomplete or "truncated" probes, with the desired full-length sequences representing, in the case of 20mer probes, approximately 10% of the total synthesis products.

For a number of reasons, the presence of truncated probe impurities has a relatively minor impact on the performance characteristics of arrays when they are used for hybridization-based sequence analysis. Firstly, the silanating agents used in this process provide an abundance of initial surface synthesis sites (>100 pmol/cm^2), so that the final "concentration" of completed probes on the support remains high. Thus, each of the 20×20 micron features on a typical array contains over 10^7 full-length oligonucleotide molecules (Fig. 2). It should be noted that there is an optimum surface probe density for maximum hybridization signal and discrimination. Thus, an increase in the synthetic yield through alternative chemistries or processes, while increasing the surface concentration of full-length probes, can actually reduce the quality of the hybridization data [10]. This is due to steric and electrostatic repulsive effects that result when oligonucleotide molecules are too densely spaced on the support. Secondly, array hybridizations are typically carried out under stringent conditions so that hybridization to significantly shorter (< approximately $n-4$) oligomers is negligible. Longer truncated sequences are of low abundance, and contribute little to the total hybridization signal in a probe feature. Furthermore, the truncated probes retain correct sequences, and any residual binding will be to the target sequences for which they were designed, albeit with slightly lower specificity. These factors, combined with the use of comparative intensity algorithms for data analysis [6], allow highly accurate sequence information to be "read" from these arrays with single-base resolution.

2.4
Future Enhancements

Improvements in Synthetic Yield. Several alternative photolabile protecting group chemistries have been described which may also be applicable to light-directed DNA array synthesis [9–15]. Some are capable of providing stepwise coupling yields in excess of 96%, and several examples are shown in Fig. 8. Achieving high synthetic yields with these alternative chemistries typically requires that a layer of solvent or catalyst be maintained over the substrate during the photodeprotection step. However, this has the drawback of adding significant complexity and cost to the manufacturing process. Furthermore, when solvent is required, the light image must be projected through the substrate from the reverse side, and image quality is somewhat degraded, thus limiting the achievable density of arrays that can be made this way. Nonetheless, certain array applications that have more stringent probe purity requirements could benefit substantially from improvements in photochemical synthesis yield. Such applications would include those in which, after hybridization, the probe-target duplexes are used as a platform for further reactions or analyses. For example, methods based on template-directed enzymatic probe extension, ligation or cleavage have been suggested as a means of improving allelic discrimination in hybridization-based mutation and polymorphism detection on arrays [16]. For this reason, we are currently developing a new generation of reagents for photolithographic synthesis that provide high synthetic yields without impacting the cost or lithographic performance of the current manufacturing process. Some of these biochemical assay formats will also require probe array synthesis to proceed in the $5'-3'$ direction so that the probes will be attached to the support at the $5'$-terminus. This can be achieved through the use of $3'$-photo-activatable $5'$-phosphoramidite building blocks [7].

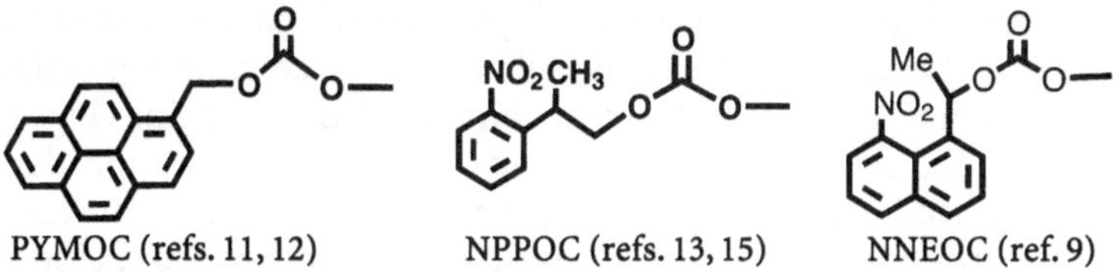

PYMOC (refs. 11, 12)   NPPOC (refs. 13, 15)   NNEOC (ref. 9)

Fig. 8. Alternate photoremovable protecting groups for photolithographic oligonucleotide synthesis

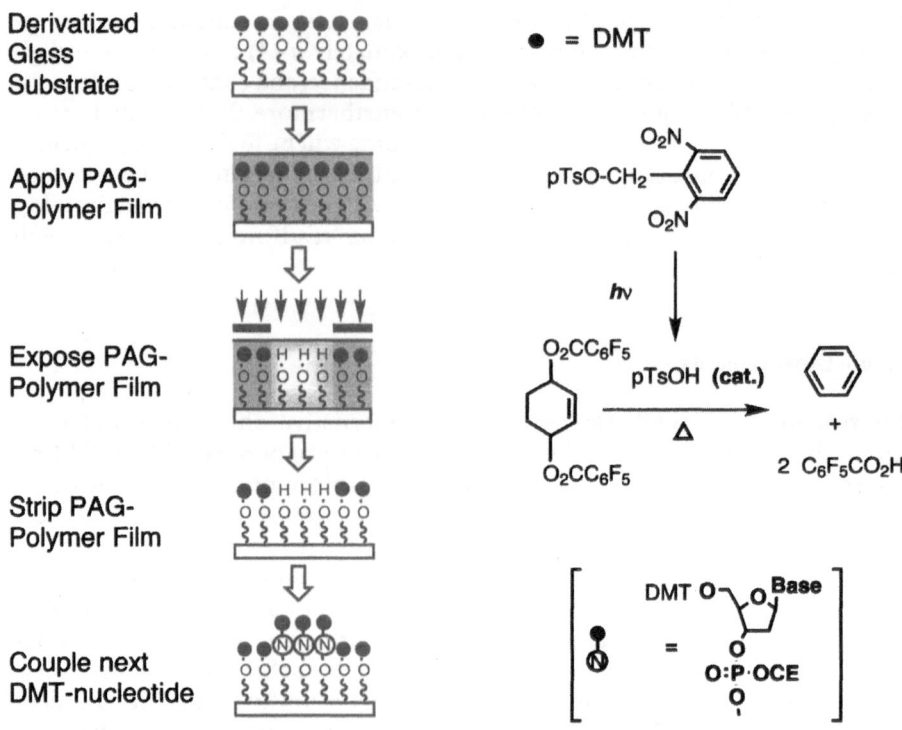

Fig 9. DNA probe array synthesis using photoacid generation in a polymer film to remove acid-labile DMT protecting groups

Improvements in Photolithographic Resolution. In order to achieve higher spatial resolution, as well as synthetic yields and photolysis rates, we have developed novel photolithographic methods for fabricating DNA arrays which exploit polymeric photo-resist films as the photoimageable component [17–20]. One of the advantages of the photo-resist approach is that it can utilize conventional 4,4′-dimethoxytrityl (DMT)-protected nucleotide monomers. These processes can also make use of chemical amplification of photo-generated catalysts to achieve higher contrast and sensitivity (shorter exposure times) than conventional photo-removable protecting groups. In this process, a polymeric thin film, containing a chemically amplified photo-acid generating (PAG) system, is applied to the glass substrate. Exposure of the film to light creates localized development of an acid catalyst in the film adjacent to the substrate surface, resulting in direct removal of DMT protecting groups from the oligonucleotide chains (Fig. 9). This process has provided stepwise synthetic yields > 98%, photolysis speeds at least an order of magnitude faster than that achievable with photo-removable protecting groups, and photolithographic resolution capability well below 10 microns. This technology will enable the production of arrays with much higher information content than is currently attainable.

Flexible Custom Array Fabrication. Another recent development has been the application of programmable digital micromirror devices, or "digital light processors" (DLPs), for photolithographic imaging, which offers a flexible approach to custom photolithographic array fabrication [21]. These devices were originally developed for digital image projection in consumer electronics products. They are essentially high-density arrays of switchable mirrors that reflect light from a source into an optical system that focuses and projects the reflected image. By using DLPs for photolithographic array synthesis, custom designs can be fabricated in a relatively short time, without the need for custom chrome-glass mask sets. This approach may offer an advantage to researchers who wish to vary designs frequently, and use only a small number of arrays of a given design. It should be noted that the standard lithographic approach using chrome-on-glass masks, which is ideal for mass production of standardized arrays, is also being adapted to the cost-effective production of smaller quantities of variable-content arrays. This is achieved through the use of high-throughput mask design and fabrication capabilities, combined with new strategies that dramatically reduce the number of masks required to synthesize custom arrays.

3
Applications

GeneChip oligonucleotide probe arrays are used to access genetic information contained in both the RNA (gene expression monitoring) and DNA (genotyping) content of biological samples. Many different GeneChip products are now available for gene expression monitoring and genotyping complex samples from a variety of organisms. The ability to simultaneously evaluate tens of thousands of different mRNA transcripts or DNA loci is transforming the nature of basic and applied research, and the range of application of DNA probe arrays is expanding at an accelerating pace. Current information on Affymetrix products and specifications is available at the website (www.affymetrix.com/products). A number of representative applications of these arrays are discussed below.

3.1
Gene Expression Monitoring

Currently, the most popular application for oligonucleotide microarrays is in monitoring cellular gene expression. Standard GeneChip arrays are encoded with public sequence information, but custom arrays are also designed from proprietary sequences. Figure 10 depicts how a gene expression array interrogates each transcript at multiple positions. This feature provides more accurate and reliable quantitative information relative to arrays that use a single probe, such as a cDNA, PCR amplicon, or synthetic oligonucleotide for each transcript. Two probes are used at each targeted position of the transcript, one complementary (perfect match probe), and one with a single base mismatch at the central position (mismatch probe). The mismatch probe is used to estimate and correct for both background and signal due to non-specific hybridization. The number of transcripts evaluated per probe array depends on chip size, the individual probe

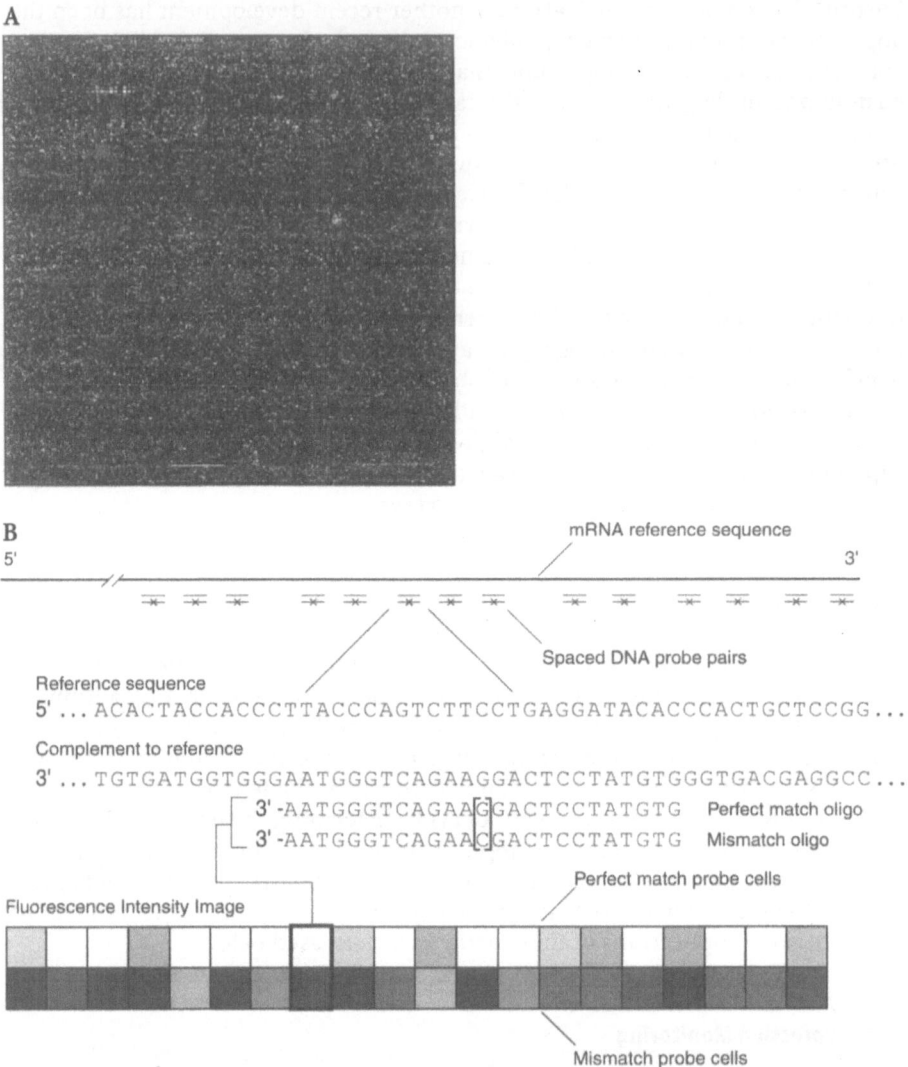

Fig. 10. Gene expression monitoring with oligonucleotide arrays. **A** An image of a hybridized 1.28 × 1.28 cm HuGeneFL array, with 20 probe pairs for each of approximately 5000 full-length human genes. **B** Probe design. To control for background and cross-hybridization, each perfect match probe is partnered with a probe of the same sequence except containing a central mismatch. Probes are usually 25mers, and are generally chosen to interrogate the 3′ regions of eukaryotic transcripts to mitigate the consequences of partially degraded mRNA. (Reprinted with permission from [6])

feature size, and the number of probes dedicated to each transcript. A standard 1.28 × 1.28 cm probe array, with individual 18 × 18 micron features, and 16 probe pairs per probe set, can interrogate over 20,000 transcripts. This number is steadily increasing as manufacturing improvements shrink the feature size, and as improved sequence information and probe selection rules allow reductions in the number of probes needed for each transcript.

Arrays are now available to examine entire transcriptomes from a variety of organisms including several bacteria, yeast, drosophila, arabidopsis, mouse, rat, and human. Instead of monitoring the expression of a small subset of selected genes, researchers can now monitor the expression of all or nearly all of the genes for these organisms simultaneously, including a large number of genes of unknown function. Numerous facets of biology and medicine are being explored using this powerful new capability. Gene function has been explored by studying changes in expression levels throughout the cell cycle [22, 23]. Genetic pathways can be examined in great detail by monitoring the downstream transcriptional effects of inducing specific genes in cell culture, and the effects of drug treatment on gene expression levels can be surveyed [24]. Expression arrays have also been used to screen thousands of genes to identify markers for human diseases such as cancer [25], muscular dystrophy [26], diabetes [27], or for aging [28, 29].

One important area of research that is benefiting greatly from GeneChip technology is cancer profiling, wherein gene expression monitoring is used to classify tumors in terms of their pathologies and responses to therapy. Understanding the variation among cancers is the key to improving their treatment. For example, a prostate tumor may be essentially harmless for 20 years in one patient, while an apparently similar tumor in another patient can be fatal within several months. One patient's lymphoma may respond well to chemotherapy while another will not. This variation of pathologies has motivated oncologists to assemble an impressive body of information to help classify tumors based on numerous histological, molecular, and clinical parameters. This has required a massive effort by thousands of highly skilled and dedicated scientists over the past few decades. It is clear, however, that there is a need for improvement in tumor classification in terms of both understanding variations among tumors, and assigning tumors to appropriate classes.

Two recent studies demonstrate the utility of GeneChip technology to cancer classification. In the first report [30], expression levels were measured in colon adenocarcinomas and normal colon tissues, and patterns differentiating the two sample types were revealed by two-way clustering analysis. It was found that ribosomal proteins were expressed at consistently higher levels in tumors than in normal tissue, thereby providing a small set of characteristic markers with large expression differences. Muscle-specific genes provided another distinguishing set of markers. These genes were expressed at a higher level in normal tissue, probably because of the tissue composition of the samples, largely epithelial tissue for the tumors and mixed tissue for the normal. In addition to identifying these two small sets of genes with relatively large expression differences, the authors report that examination of large sets of genes with even small differences in expression could reliably classify a sample as tumor or normal. Although the confidence level of an individual gene might be low, extensive expression analysis using thousands of genes reveals subtle, systematic differences with high confidence. Such an expression profile database constructed from samples of known types would be useful for class prediction, that is, classifying additional unknown samples.

In a second report [31], class predictions were tested directly using a database built from two different types of leukemia. The current clinical tests used to distinguish ALL (acute lymphoblastic leukemia) from AML (acute myeloid leukemia), while useful, are imperfect and painstaking. Proper identification of these cancer types is critical because the treatments they require are quite different. In the first set of experiments, the authors examined 27 ALL and 11 AML samples to create a gene expression profile database. Statistical analysis identified about 1100 genes (of the 6817 examined) that correlated well with the ALL-AML class distinction. Many of the most highly correlated genes had been previously implicated in cancer. The 50 most informative genes were used to accurately cross-validate 36 of 38 of the original samples (the other 2 calls were uncertain), and were also good predictors for an independent set of leukemia samples (29 of 34 were called correctly; the other 5 were uncertain). The test on independent samples is especially convincing given the more varied nature of the samples: two different RNA isolation methods were used by different laboratories; some samples were isolated from peripheral blood, some from bone marrow; some AML patients were adults, and some were children. Thus, the panel of the 50 most informative genes served as a strong class prediction set, and the authors reported similar results using 10–200 genes. The authors point out, however, that for other sample groups the number of genes needed for accurate class prediction may vary, and that no single gene correlated completely with class type.

Golub et al. [31] went on to address the issue of class discovery, that is, the identification of new cancer classes. The statistical tool called self-organizing maps was employed to determine whether the original data set could be subdivided beyond the ALL-AML categories. The AML samples again clustered together, but the ALL samples were now split into two groups that were subsequently shown by immunotyping to be of B-cell or T-cell origin. Although this ALL subdivision was previously known, the clustering analysis would have discovered it even if it had not been known.

Expression profiling coupled with appropriate statistical analysis holds promise not only in cancer classification, but also by extension to many other areas of disease research and management. Transcription profile databases may be assembled from samples that differ by tissue source, disease state or progression, morbidity/mortality, response to drugs and other treatments, and countless other variables. New patterns may be revealed and disease classes refined or discovered. Patients may be more finely stratified in clinical trials so that the success of treatments can be better judged, and the expectation is that the diagnosis and treatment of disease will improve substantially as a result.

3.2
Genotypic Analysis

As the human genome project finishes the first complete blueprint of the human genome, there is tremendous interest in identifying the variations in DNA sequences between individuals and relating these variations to phenotypes. It is of particular interest to understand how subtle sequence differences are associated with disease. Oligonucleotide arrays are well suited to probe these variations, particularly single-nucleotide substitutions and, to a lesser degree, short deletions and insertions.

Oligonucleotide arrays are currently used primarily for two types of genotyping analysis. *Arrays for mutation or variant detection* (Fig. 11) are used to screen sets of contiguous sequence for single-nucleotide differences. Given a reference sequence, the basic design of genotyping arrays is quite simple: four probes, varying only in the central position and each containing the reference sequence at all other positions, are made to interrogate each nucleotide of the reference sequence. The target sequence hybridizes most strongly to its perfect complement on the array, which in most cases will be the probe corresponding to the reference sequence, but, in the case of a nucleotide substitution, this will be one of the other three probes. Up to 50 kb of sequence can thus be determined with 200,000 probes, or 400,000 probes if both DNA strands are interrogated. Impending decreases in array feature size (see Sect. 2.4) will extend this capacity further. Mutation detection arrays have been used to analyze the entire 16.5 kb sequence of mitochondrial DNA samples [32], the 9.2 kb coding sequence of the ATM gene [33], BRCA1 coding mutations [34, 35], p53 mutations [36, 37], cytochrome P450 variants involved in drug metabolism [38], and HIV sequence variants [39, 40], among others. The last three arrays are commercial products currently available from Affymetrix.

The other main type of genotyping performed with oligonucleotide arrays is *SNP analysis*, that is, the genotyping of biallelic single-nucleotide polymor-

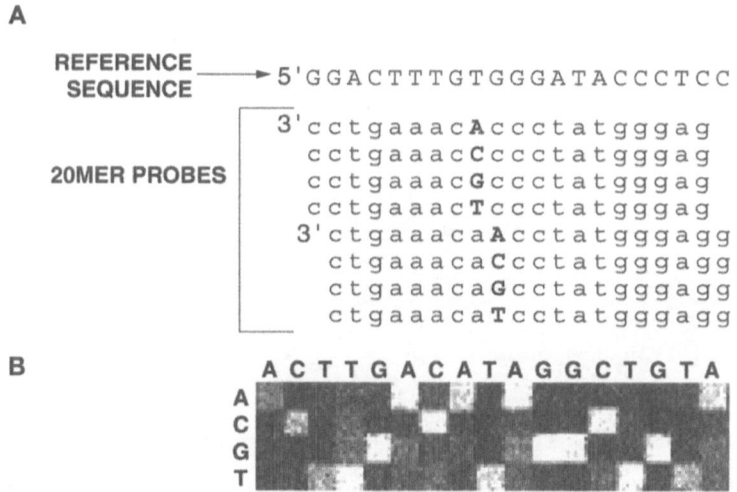

Fig. 11. Resequencing array for sequence variation detection. **A** Each base of a given reference sequence is represented by four probes, usually 20mers, that are identical to each other with the exception of a single centrally located substitution (*bold*). Shown are probe sets targeted to two adjacent positions of the reference sequence. **B** The target sequence is determined by hybridization intensities, with the probe complementary to the target providing the strongest signal. (Reprinted with permission from Warrington JA et al (2000) In: Microarray biochip technology. Biotechniques Books, p 122)

phisms. Because SNPs are the most common source of variation between individuals, they serve not only as landmarks to create dense genome maps, but also as markers for linkage and loss of heterozygosity studies. Large numbers of publicly available SNPs – over one million to date – have been found using dideoxy sequencing as well as mutation detection arrays [41–43].

In addition to mutation detection arrays, at least two other types of oligonucleotide arrays can be used for SNP analysis. The "HuSNP" assay allows nearly 1500 SNP-containing regions of the human genome to be amplified in just 24 multiplex PCRs and then hybridized to a single HuSNP array. The SNPs cover all 22 autosomes and the X chromosome. Many of the 1500 SNPs were discovered using mutation detection arrays [41]. The probe strategy for a SNP array is shown in Fig. 12. The probes for each SNP on the HuSNP array interrogate not only the two alleles of the SNP position, but also three or four positions flanking the SNP; the redundant data are of higher quality for the same reasons that the use of multiple probes improves gene expression monitoring array data. SNP arrays have also proven useful for loss of heterozygosity studies [44, 45].

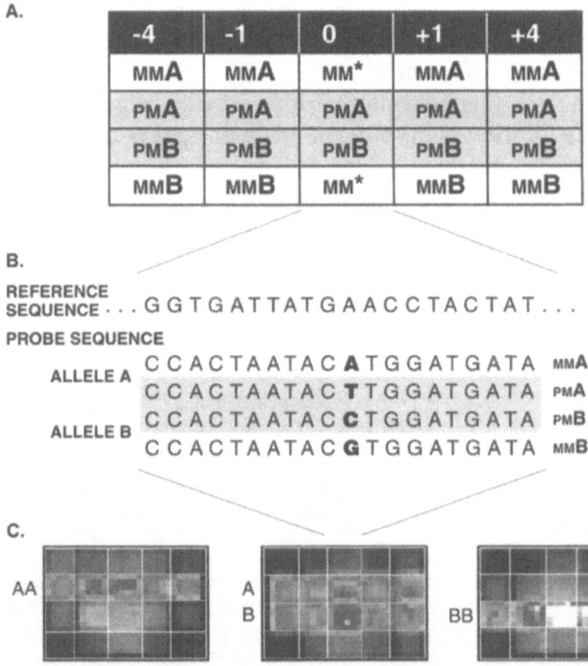

Fig. 12. HuSNP array design. **A** A known biallelic polymorphism at position 0 is interrogated by a block of four or five probe sets (five in this example). Each probe set consists of four probes, a perfect match and a mismatch to allele A, and a perfect match and a mismatch to allele B. One probe set in a block is centered directly over the polymorphism "0"), and others are centered upstream (−4, −1) and downstream (+1, +4). **B** The sequences of the probe set centered over the polymorphism is shown. **C** Sample images of blocks showing homozygous *A*, heterozygous *A/B*, or homozygous *B* at the same SNP site. (Reprinted with permission from Warrington JA et al (2000) In: Microarray biochip technology. Biotechniques Books, p 122)

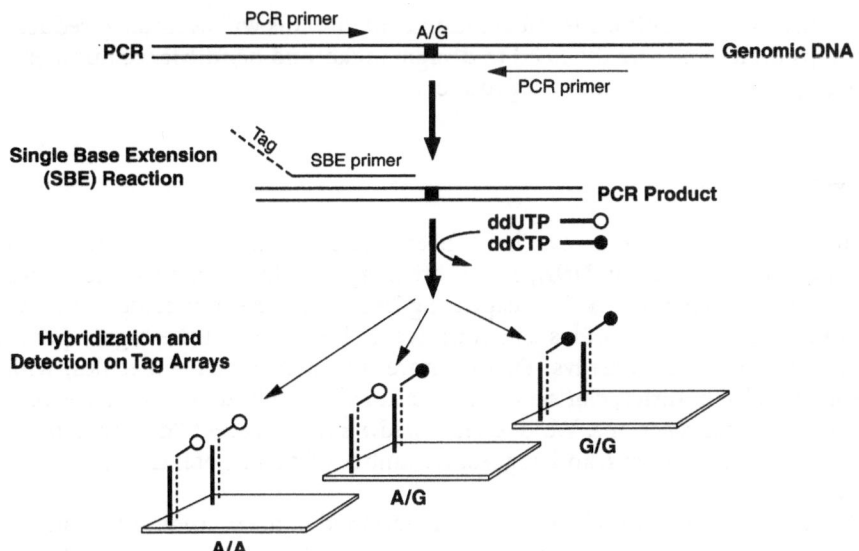

Fig. 13. Schematic of the single-base extension assay applied to Tag probe arrays. Regions containing known SNP sites (A or G in this example) are first amplified by PCR. The PCR product serves as the template for an extension reaction from a chimeric primer consisting of a 5' tag sequence and a 3' sequence that abuts the polymorphic site. The two dideoxy-NTPs that could be incorporated are labeled with different flurophors; in this example, ddUTP is incorporated in the case of the A allele, and ddCTP for the G allele. Multiple SBE reactions can be done in a single tube. The tag sequence, unique for each SNP, directs the extension products to a particular address on the Tag probe array. The proportion of a fluorophor at an address reflects the abundance of the corresponding allele in the original DNA. (Reprinted with permission from [45])

Although it is anticipated that the HuSNP assay will be appropriate for many applications, a more generic alternative is available in the form of the GenFlex array. For this array, 2000 20mer "tag" probe sequences were selected on the basis of uniform hybridization properties and sequence specificity. The array includes three control probes for each tag (a complementary probe and single-base mismatch probes for both the tag and its complement). One way to use the GenFlex array for SNP analysis is illustrated below (Fig. 13). In this example, a single-base extension reaction is used, in which a primer abutting the SNP is extended by one base in the presence of the two possible dideoxy-NTPs, each of which is labeled with a different fluorophor. Since each target-specific primer carries a different tag, the identity of each SNP is determined by hybridization of the single-base extension product to the corresponding tag probe in the GenFlex array [46]. The flexibility of the GenFlex approach lies in the freedom to partner any primer with any tag, a feature that enables other applications as well (Sect. 3.3).

Unlike yeast or bacteria, the size and complexity of the human genome currently necessitates locus-specific amplification for these genotyping applications. Without amplification, the concentration of each target is simply too low. We are developing more general arrays and procedures to reduce sequence complexity while maintaining sufficient information content. This will eventually reduce or perhaps even eliminate the need to design, make, and handle large numbers of locus-specific primers and PCR products.

3.3
Other Applications

While oligonucleotide arrays have been used primarily to determine the composition of RNA or DNA, many other applications are possible as well. Any methodology that involves capturing large numbers of molecules that will hybridize to oligonucleotides can conceivably benefit from the highly parallel nature of these microarrays. Furthermore, the hybridized molecules, which are essentially libraries, can serve as a platform for subsequent analyses based on biochemical reactions. We describe below several recent "non-traditional" uses of GeneChip arrays, and suggest a number of other potential applications as well.

Tag arrays, such as the GenFlex array mentioned in the preceding section, have been used as "molecular bar-code" detectors [47 – 49]. In these experiments, mixtures of multiple yeast strains – each carrying a unique tag in its genome and having a different gene deleted – were subjected to a test such as drug treatment or growth in minimal medium, and then tag probe arrays were used to determine the proportion of each strain in the surviving population. As in gene expression and genotyping applications, the molecular bar-coding strategy takes advantage of the ability of probe arrays to selectively bind many different sequences in a complex mixture simultaneously. Parallel processing is not only faster and easier – it also minimizes the effect of variations in sample handling, thereby increasing the accuracy and precision of the measurements.

It is also envisioned that tag probe arrays will be useful for proteomics and other protein screening applications. For example, by attaching a different oligonucleotide sequence tag to each member of a group of proteins to be analyzed, hybridization would allow them to be arrayed in discrete locations on a chip for parallel screening. Proteins of interest would be identified by their position on the array. In one possible approach (Fig. 14), the tag is attached to the protein genetically by linking the tag to the mRNA and then translating the protein in such a manner that the protein remains associated with the mRNA, as is done in ribosome display to create and capture high affinity antibodies [50]. The protein-mRNA-tag complex is hybridized to the tag probe array, and screened for protein activity on the array. It is also conceivable that the proteins could be translated on the array, after hybridization. Genes of interest are recovered, either directly from the array or from another aliquot of the mRNA library, by PCR using the tag sequence for one primer and a common 3′ end sequence as the other primer. One use for such a system would be in directed evolution projects in which large gene libraries are made by cloning into cells, usually bacteria or yeast, followed by propagating and screening each clone individually for production of a protein with new or improved properties. The tag system would not only eliminate the need to transform and handle individual clones, but would also allow highly parallel screening because thousands of variants could be assayed simul-

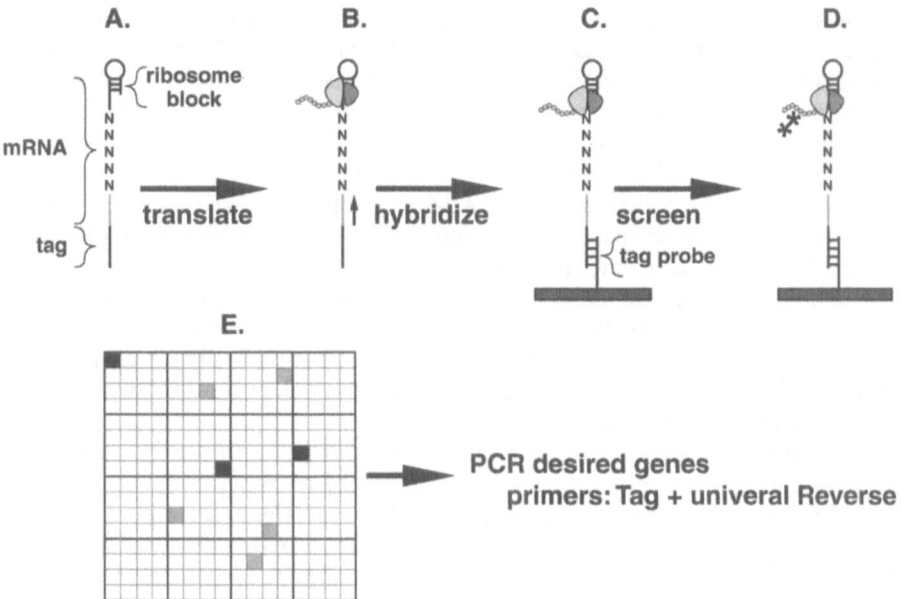

Fig. 14. Using Tag probe arrays to screen protein activity. To a protein-encoding mRNA a 5′ tag sequence and a 3′ ribosome-blocking sequence are attached (**A**). In a pool of such molecules, such as a randomly mutated gene library, each mRNA is paired with a unique tag and all have the same 3′ sequence. Following in-vitro translation either on a microarray or in a test tube, the nascent protein remains attached to the mRNA (**B**). During hybridization the tag directs each mRNA-protein to a particular address on the Tag probe array (**C**), where all the proteins are screened simultaneously for activity (**D**). Appropriate detection methods identify proteins of interest (**E**, *black and/or shaded blocks*). Finally, the corresponding genes can be captured by PCR of the mRNA pool using a universal reverse primer and each identified Tag sequence as a forward primer

taneously on the same array. Another use for the tag system would be to screen (poly)peptides made from existing mRNA molecules for properties such as drug binding. For example, all the mRNAs from a pathogenic bacterial strain could be converted to tagged proteins, which could then be screened for the ability to bind antibiotic candidates. The RNA molecules themselves could also be screened, as some drugs act directly on RNA. It is also conceivable that the oligonucleotide tag could be added directly to proteins, a method that might be useful in cases in which clones are already separated and one wishes to use the tag probe array only for parallel screening.

Researchers have also found a variety of novel uses for GeneChip arrays that were originally designed for gene expression monitoring. For example, Cho and co-workers [51] carried out a yeast two-hybrid study of *S. cerevisiae* proteins, mixing DNA from positive clones and hybridizing them to a yeast expression array, enabling them to identify the clones much more efficiently than if they had sequenced the clones by traditional methods. Winzeler et al. [52] used yeast expression arrays to identify more than 3700 biallelic genomic variations between two yeast strains and then used the markers to simultaneously map five loci with high resolution (11 – 64 kb). Deletions in yeast [53] and in a clinical *M. tuberculosis* strain [54] were identified by similar techniques, a potentially important application given the propensity of some pathogens to avoid drug and vaccination treatments by deleting segments of their genomes. The use of multiple probes for each gene, a characteristic feature of GeneChip expression arrays, was critical to the high degree of resolution that was achieved in these experiments.

Milner et al. [55] reported using oligonucleotide arrays to survey oligonucleotide binding to a specific mRNA. A prevalent approach in anti-gene therapeutics involves knocking out malfunctioning genes through RNase H-mediated degradation of the mRNA, induced by duplex formation with antisense oligonucleotides. Presumably, the ability to predict, or at least empirically select, oligonucleotides which hybridize best to a given mRNA would be useful in the development of anti-gene therapeutics.

Gene expression arrays have also served as platforms for the analysis of splice variation in organisms with introns, and for mapping transcriptional boundaries [56]. Also, samples can be preselected for certain properties before hybridization, and at least two examples of this have been reported. In one case, cellular RNA [57], and in the other cellular DNA [58], were mixed with specific proteins, and complexes were purified by immunoprecipitation. Hybridization of the nucleic acids from the purified complexes revealed specific associations with the proteins. These two elegant experiments were carried out using arrays of spotted PCR products, but, again, one would expect that data of even higher resolution would be achievable using the multi-sequence probe sets present on GeneChip expression monitoring arrays.

One could conceivably treat hybridized expression arrays with RNase H, and assay for activity directly by following the loss of signal. This type of approach could be useful for revealing potential RNase H "hotspots" in mRNAs. A number of other powerful, but as yet under-utilized, applications also use arrayed probe-target duplexes as a platform for further reactions or analyses [59]. For example, Bulyk, Church and co-workers [60] created arrays of four base-pair restriction enzyme recognition sites and demonstrated activity with the appropriate enzymes, including *dam* methylase. Studies such as this provide further demonstration that arrays of double-stranded probes offer a promising platform for studying DNA-protein interactions. Methods based on template-directed enzymatic probe extension, ligation or cleavage are also being investigated as a means of improving allelic discrimination in hybridization-based mutation and polymorphism detection on arrays [16]. It is expected that hybridization-based biochemical assays on DNA microarrays will become increasingly commonplace in the coming years, especially in the area of high-throughput genotyping applications.

4
References

1. Phimister B (ed) (1999) Nat Genet Suppl 21:1
2. Schena R, Davis RW (2000) In: Schena M (ed) Microarray biochip technology. BioTechniques Books, Natick, MA, p 1
3. Fodor SPA, Read JL, Pirrung MC, Stryer LT, Lu A, Solas D (1991) Science 251:767
4. Pease AC, Solas D, Sullivan EJ, Cronin MT, Holmes CP, Fodor SPA (1994) Proc Natl Acad Sci USA 91:5022
5. McGall GH, Barone AD, Diggelmann M, Fodor SPA, Gentalen E, Ngo N (1997) J Am Chem Soc 119:5081
6. Lipshutz R, Fodor SPA, Gingeras TR, Lockhart DJ (1999) Nat Genet Suppl 21:20
7. McGall GH, Fidanza JA (2001) In: Rampal JB (ed) Methods in molecular biology. DNA arrays methods and protocols. Humana Press, Inc., Totowa, NJ, p 71
8. McGall GH, Barone AD, Diggelmann M (1999) Eur Pat Appl EP 967,217
9. Barone AD, Beecher JE, Bury P, Chen C, Doede T, Fidanza JA, McGall GH (2001) Nucleosides Nucleotides 20:525
10. Unpublished data
11. McGall GH (1997) In: Hori W (ed) Biochip arrays. IBC Library Series, Southboro, MA, p 2.1
12. McGall GH, Nam NQ, Rava R (2000) US Patent 6,147,205
13. Hasan A, Stengele K-P, Giegrich H, Cornwell P, Isham KR, Sachleben R, Pfleiderer W, Foote RS (1997) Tetrahedron 53:4247
14. Pirrung MC, Fallon L, McGall G (1998) J Org Chem 63:241
15. Beier M, Hoheisel JD (2000) Nucleic Acids Res 28:e11
16. Tonisson N, Kurg A, Lohmussaar E, Metspalu A (2000) In: Schena M (ed) Microarray biochip technology. BioTechniques Books, Natick, MA, p 247
17. McGall G, Labadie J, Brock P, Wallraff G, Nguyen T, Hinsberg W (1996) Proc Natl Acad Sci USA 93:13555
18. Wallraff G, Labadie J, Brock P, Nguyen T, Huynh T, Hinsberg W, McGall G (1997) Chemtech, February:22
19. Beecher JE, McGall GH, Goldberg MJ (1997) Preprints Am Chem Soc Div Polym Mater Sci Eng 76:597
20. Beecher JE, McGall GH, Goldberg MJ (1997) Preprints Am Chem Soc Div Polym Mater Sci Eng 77:394
21. Singh-Gasson S, Green RD, Yue Y, Nelson C, Blattner F, Sussman MR, Cerrina F (1999) Nat Biotechnol 17:974
22. Cho RJ, Campbell MJ, Winzeler EA, Steinmetz L, Conway A, Wodicka L, Wolfsberg TG, Gabrielian AE, Landsman D, Lockhart DJ, Davis RW (1998) Mol Cell 2:65
23. Cho RJ, Huang M, Dong H, Steinmetz L, Sapinoso L, Hampton G, Elledge SJ, Davis RW, Lockhart DJ, Campbell MJ (2001) Nat Genet 27:48
24. Debouck C, Goodfellow PN (1999) Nat Genet Suppl 21:48
25. Liotta L, Petricoin E (2000) Nat Rev Genet 1:48
26. Chen YW, Zhao P, Borup R, Hoffman EP (2000) J Cell Biol 151:1321
27. Wilson SB, Kent SC, Horton HF, Hill AA, Bollyky PL, Hafler DA, Strominger JL, Byrne MC (2000) Proc Natl Acad Sci USA 97:7411
28. Lee CK, Klopp RG, Weindruch R, Prolla TA (1999) Science 285:1390
29. Ly DH, Lockhart DJ, Lerner RA, Schultz PG (2000) Science 287:2486
30. Alon U, Barkai N, Notterman DA, Gish K, Ybarra S, Mack D, Levine AJ (1999) Proc Natl Acad Sci USA 96:6745
31. Golub TR, Slonim DK, Tamayo P, Huard C, Gaasenbeek M, Mesirov JP, Coller H, Loh ML, Downing JR, Caligiuri MA, Bloomfield CD, Lander ES (1999) Science 286:531
32. Chee M, Yang R, Hubbell E, Berno A, Huang XC, Stern D, Winkler J, Lockhart DJ, Morris MS, Fodor SP (1996) Science 274:610

33. Hacia JG, Sun B, Hunt N, Edgemon K, Mosbrook D, Robbins C, Fodor SP, Tagle DA, Collins FS (1998) Genome Res 8:1245
34. Hacia JG, Brody LC, Chee MS, Fodor SP, Collins FS (1996) Nat Genet 14:441
35. Hacia JG (1999) Nat Genet 21:42
36. Ahrendt SA, Halachmi S, Chow JT, Wu L, Halachmi N, Yang SC, Wehage S, Jen J, Sidransky D (1999) Proc Natl Acad Sci USA 96:7382
37. Wikman FP, Lu ML, Thykjaer T, Olesen SH, Andersen LD, Cordon-Cardo C, Orntoft TF (2000) Clin Chem 46:1555
38. Liu WW, Webster T, Aggarwal A, Pho M, Cronin M, Ryder T (1997) Am J Hum Genet 61:1494
39. Kozal MJ, Shah N, Shen N, Yang R, Fucini R, Merigan TC, Richman DD, Morris D, Hubbell E, Chee M, Gingeras TR (1996) Nat Med 2:753
40. Gunthard HF, Wong JK, Ignacio CC, Havlir DV, Richman DD (1998) AIDS Res Hum Retroviruses 14:869
41. Wang DG, Fan JB, Siao CJ, Berno A, Young P, Sapolsky R, Ghandour G, Perkins N, Winchester E, Spencer J, Kruglyak L, Stein L, Hsie L, Topaloglou T, Hubbell E, Robinson E, Mittmann M, Morris MS, Shen N, Kilburn D, Rioux J, Nusbaum C, Rozen S, Hudson TJ, Lander ES, et al (1998) Science 280:1077
42. Cargill M, Altshuler D, Ireland J, Sklar P, Ardlie K, Patil N, Shaw N, Lane CR, Lim EP, Kalyanaraman N, Nemesh J, Ziaugra L, Friedland L, Rolfe A, Warrington J, Lipshutz R, Daley GQ, Lander ES (1999) Nat Genet 22:231
43. Lindblad-Toh K, Winchester E, Daly MJ, Wang DG, Hirschhorn JN, Laviolette JP, Ardlie K, Reich DE, Robinson E, Sklar P, Shah N, Thomas D, Fan JB, Gingeras T, Warrington J, Patil N, Hudson TJ, Lander ES (2000) Nat Genet 24:381
44. Lindblad-Toh K, Tanenbaum DM, Daly MJ, Winchester E, Lui WO, Villapakkam A, Stanton SE, Larsson C, Hudson TJ, Johnson BE, Lander ES, Meyerson M (2000) Nat Biotechnol 18:1001
45. Mei R, Galipeau PC, Prass C, Berno A, Ghandour G, Patil N, Wolff RK, Chee MS, Reid BJ, Lockhart DJ (2000) Genome Res 10:1126
46. Fan JB, Chen X, Halushka MK, Berno A, Huang X, Ryder T, Lipshutz RJ, Lockhart DJ, Chakravarti A (2000) Genome Res 10:853
47. Shoemaker DD, Lashkari DA, Morris D, Mittmann M, Davis RW (1996) Nat Genet 14:450
48. Giaever G, Shoemaker DD, Jones TW, Liang H, Winzeler EA, Astromoff A, Davis RW (1999) Nat Genet 21:278
49. Winzeler EA, Shoemaker DD, Astromoff A, Liang H, Anderson K, Andre B, Bangham R, Benito R, Boeke JD, Bussey H, Chu AM, Connelly C, Davis K, Dietrich F, Dow SW, El Bakkoury M, Foury F, Friend SH, Gentalen E, Giaever G, Hegemann JH, Jones T, Laub M, Liao H, Davis RW, et al (1999) Science 285:901
50. Hanes J, Jermutus L, Pluckthun A (2000) Methods Enzymol 328:404
51. Cho RJ, Fromont-Racine M, Wodicka L, Feierbach B, Stearns T, Legrain P, Lockhart DJ, Davis RW (1998) Proc Natl Acad Sci USA 95:3752
52. Winzeler EA, Richards DR, Conway AR, Goldstein AL, Kalman S, McCullough MJ, McCusker JH, Stevens DA, Wodicka L, Lockhart DJ, Davis RW (1998) Science 281:1194
53. Winzeler EA, Lee B, McCusker JH, Davis RW (1999) Parasitology 118:S73
54. Salamon H, Kato-Maeda M, Small PM, Drenkow J, Gingeras TR (2000) Genome Res 10:2044
55. Milner N, Mir KU, Southern EM (1997) Nat Biotechnol 15:537
56. Wassarman KM, Repoila F, Rosenow C, Storz G, Gottesman S (2001) Genes Dev 15:1637
57. Takizawa PA, DeRisi JL, Wilhelm JE, Vale RD (2000) Science 290:341
58. Ren B, Robert F, Wyrick JJ, Aparicio O, Jennings EG, Simon I, Zeitlinger J, Schreiber J, Hannett N, Kanin E, Volkert TL, Wilson CJ, Bell SP, Young RA (2000) Science 290:2306
59. Gunderson KL, Huang XC, Morris MS, Lipshutz RJ, Lockhart DJ, Chee MS (1998) Genome Res 8:1142
60. Bulyk ML, Gentalen E, Lockhart DJ, Church GM (1999) Nat Biotechnol 17:573

Received: June 2001

Oligonucleotide Scanning Arrays: Application to High-Throughput Screening for Effective Antisense Reagents and the Study of Nucleic Acid Interactions

M. Sohail · E. M. Southern

University of Oxford, Department of Biochemistry, South Parks Road, Oxford OX1 3QU, England, UK. *E-mail: msohail@bioch.ox.ac.uk*

Oligonucleotide scanning arrays are useful tools in the study of nucleic acid interaction. Such arrays of oligonucleotides, corresponding to a full set of complements of a known sequence, can be readily made in a single series of coupling reactions, adding each nucleotide in the complement of the target sequence in turn. The synthesis is carried out on the surface of a solid substrate such as glass or polypropylene that has been modified to allow nucleotide coupling. A mask is used to apply the DNA synthesis reagents in a defined area and is moved by a fixed step size after each coupling reaction so that consecutive couplings overlap a portion of the previous one. The size of the mask and the displacement at each coupling determine the length of the oligonucleotides.

A radiolabeled or fluorescently tagged target sequence is hybridised to a scanning array and its interaction with the complementary oligonucleotides is displayed as a hybridisation signal. It is thus possible to determine the exact sequence and lengths of large numbers of interacting sequences in a single hybridisation experiment. The array image is analysed using a computer program (xvseq) that calculates quantitative measurements of the binding strengths. We have found scanning arrays a useful tool not only to find effective antisense reagents, but also to study RNA folding and the mechanisms of heteroduplex formation. In this article, we discuss the format of these arrays, the technology used to fabricate and to read them, and their applications.

1	**Introduction**	44
2	**The Format of Scanning Arrays**	44
3	**Making and Reading Scanning Arrays**	47
3.1	Fabrication	47
3.1.1	Solid Supports for Making Arrays	47
3.1.2	Masks for Making Arrays: Materials and Machining	48
3.1.3	Making Scanning Arrays on an ABI DNA/RNA Synthesiser	48
3.1.4	Deprotection of Arrays	50
3.2	Hybridisation of a Labeled Target to Scanning Arrays	51
3.3	Reading a Scanning Array Image	52
4	**Applications**	53
4.1	High-Throughput Screening of Antisense Reagents	53
4.2	The Study of Nucleic Acid Folding and Heteroduplex Formation	54

Advances in Biochemical Engineering/
Biotechnology, Vol. 77
Managing Editor: T. Scheper
© Springer-Verlag Berlin Heidelberg 2002

4.2.1   RNA Folding 54
4.2.2   Heteroduplex Formation 55
4.2.3   A Study of the Secondary Structure of Nucleic Acids with
        Modified Bases 55

5   Concluding Remarks 56

6   References 56

1
Introduction

Scanning arrays are made by combinatorial in situ synthesis of a large number
of oligonucleotides on a solid support in a spatially addressable fashion. They
comprise sets of oligonucleotides of various lengths in which a series of oligonucleotides, complementary to the target RNA or DNA, is made by sequential coupling of nucleotides to a solid substrate (such as glass or polypropylene) that has
been modified to allow oligonucleotide synthesis. The set of oligonucleotides
covers every nucleotide position in the region of interest of the complement of
the target. The resultant array allows an exhaustive and parallel analysis of the interactions of a complete set of oligonucleotides complementary to the target under investigation in a single hybridisation experiment (see flow chart in Fig. 1).

2
The Format of Scanning Arrays

The process of array fabrication allows the synthesis of oligonucleotides that differ from the adjacent members in the set by one or two nucleotides and they are
in the order in which they occur in the complement of the target sequence. If the
mask used to make arrays is kept in the same place during all the synthesis cycles, the result will be a complete complement of the target sequence. However,
if the mask is moved along the surface after each nucleotide coupling, the result
is a series of oligonucleotides, each one complementary to a region of the target
sequence. The displacement of the mask is such that the mask overlaps a portion
of its footprint at the previous position. The oligonucleotides range in size from
monomers to a certain maximum, generally around 20 – 25 residues. The longest
in the set of the oligonucleotides is determined by the size of the mask used to
retain DNA synthesis reagents and the displacement of the mask after each nucleotide coupling (Fig. 2).

We generally use either diamond-shaped or circular masks to make scanning
arrays. For a diamond-shaped mask, the length of the longest oligonucleotide is
the ratio of the diagonal of the mask to the displacement and for a circular mask
it is the ratio of the diameter of the circle to the displacement at each nucleotide
coupling. For example, with a diamond-shaped mask of 30 mm diagonal, a displacement of 2 mm at each addition will result in 15-mers along the centre line,
while a displacement of 1.5 mm will result in 20-mers. The cells created by over-

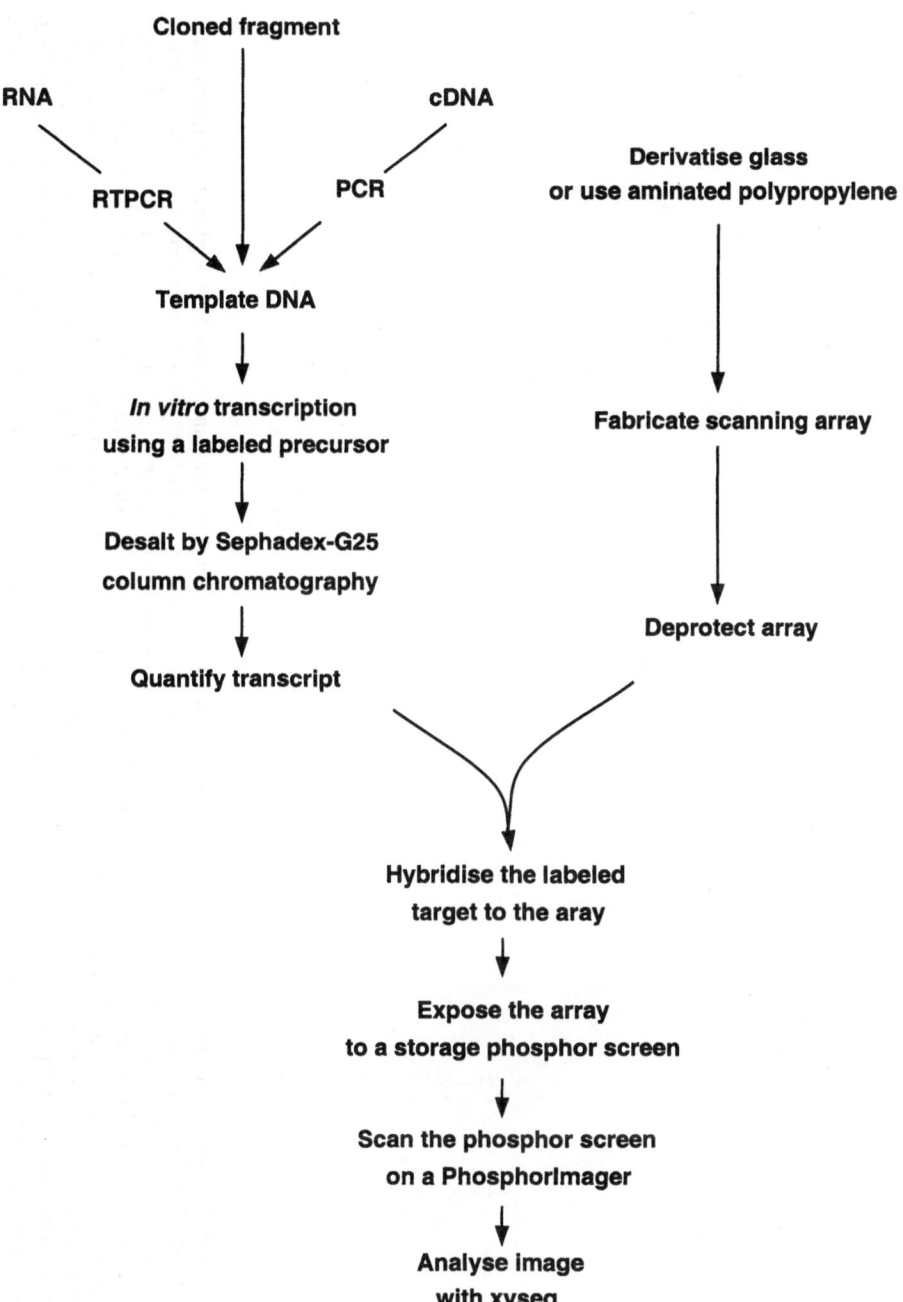

Fig. 1. Flow chart summarising the process of scanning array farication, hybridisation and imaging

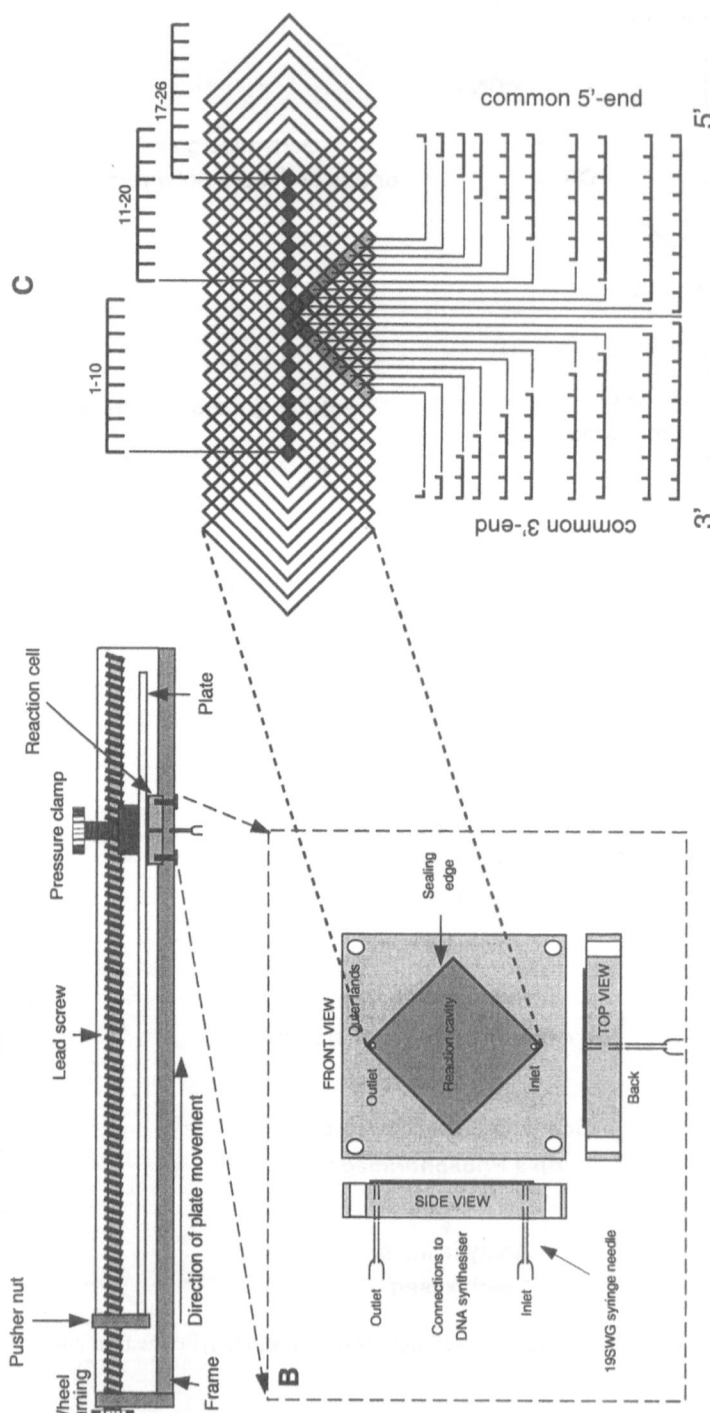

Fig. 2. **A** Apparatus to make scanning arrays by hand. The mask shown in **B** is fixed on the frame and the frame (made from angled aluminium) is fixed in front of a DNA synthesiser. The lead screw is fitted with a pusher nut that drives the plate across the front of the mask by driving the turning wheel. The pressure clamp is fixed to the back of the frame with a polyethylene cushion mounted on its pressure pad. In the automated version of the array maker, both the turning wheel and the pressure clamp have been replaced with computer-controlled motors. **B** A diamond-shaped mask. The mask consists of a block (metal or PTFE) with a diamond-shaped sealing edge. Two holes, drilled just inside of the top and the bottom corners (marked *inlet* and *outlet*) of the diamond shape, can take shortened 19SWG syringe needles and are used to connect the mask to the reagents supply of the DNA synthesiser. **C** Format of scanning arrays. A template grid of overlapping diamonds is shown. The longest oligonucleotides in the set are 10-mers (*closed diamonds*) located along the centre line of the grid. The positions of three 10-mers (1–10, 11–20 and 17–26) are marked on the top of the grid and the lengths of the oligonucleotides from monomers to 10-mers with common 3′- and common 5′-ends at the bottom. The arrays are symmetrical above and below the centre line allowing duplicate hybridisation measurements. Each smaller diamond-shaped cell is occupied by a single oligonucleotide sequence

lapping footprints of the diamond-shaped mask are small diamonds but the cells created by a circular mask differ in shape: along the centre line they are lenticular but off this line they form a four-cornered "spearhead" that diminishes in size towards the edge. Each of these smaller cells contains a single oligonucleotide sequence that differs from oligonucleotides in the neighbouring cells by one or two nucleotides, depending on its position (see Fig. 2). The longest oligonucleotides are found along the centre line of the array and monomers are located along the edges. In between are all lengths ranging from monomers to the longest in the set. The number of oligonucleotides of a particular length can be calculated by a simple formula, $N - s + 1$, where s is the length of the oligonucleotides and N is the total number of nucleotide couplings used to make the array. For example, if a 100 nt long sequence is covered in 100 coupling steps (with a diamond-shaped mask of 30 mm diagonal and 1.5 mm step size), there will be 100 monomers, 96 5-mers, 91 10-mers, and 81 20-mers. The last 19 positions in sequence will be represented by shorter oligonucleotides only, from 19-mers to monomers in this case. Therefore, to make 100 20-mers in the array, an additional 19 nt synthesis steps must to be added at the end (total coupling steps $= N + s - 1$). The arrays as fabricated are symmetrical above and below the centre line, thus each oligonucleotide is represented twice on the array allowing for duplicate hybridisation measurement providing an internal hybridisation control. Additionally, each half can be hybridised separately. In this way, one can measure the hybridisation of two types of targets to the same set of oligonucleotides under identical conditions, or study the interaction of the same target with oligonucleotides under different hybridisation conditions, allowing a direct comparison between two experiments.

3
Making and Reading Scanning Arrays

3.1
Fabrication

3.1.1
Solid Supports for Making Arrays

A flat and impermeable surface is required for in situ synthesis of oligonucleotides and both glass and polypropylene meet these criteria.

Glass has a number of favourable properties including its easy availability, smooth surface, physical and chemical stability, and compatibility with the use of radiolabeled and fluorescently labeled targets. We have found that 3 mm window glass is a good material for array fabrication. To make the glass surface ready for chemical reaction with nucleotide precursors (standard nucleotide CE phosphoramidites), it is chemically derivatised with a 20-atom aliphatic chain to produce free hydroxyl groups on its surface [1]. Glass is first treated with a solution of glycidoxypropyl trimethoxysilane that reacts with the silanol groups on its surface to produce an epoxide group. This is then reacted with a polyethylene glycol leaving a linker/spacer molecule terminating in a hydroxyl group that is a substrate for reaction with nucleotide phosphoramidite [1]. Alternatively, the glass can be derivatised by exposing it to a vapour phase of trimethoxysilane in a vacuum furnace at approximately 140–145 °C for several hours. This creates a monolayer of the epoxide group on the surface that can then be reacted with polyethylene glycol.

Polypropylene also has favourable physical and chemical properties. Its surface is aminated by plasma discharge in the presence of anhydrous ammonia [2]. The free amine groups ($-NH_2$) thus produced on its surface are also a substrate for oligonucleotide synthesis with standard phosphoramidite chemistry.

3.1.2
Masks for Making Arrays: Materials and Machining

Both circular and diamond-shaped masks can be made from stainless steel, aluminium or PTFE (Teflon): the process of making them may differ with the type of the material used and the shape of the mask. Circular masks are made using a centre lathe and can be made from both metals and PTFE. Diamond-shaped masks are rather difficult to make with PTFE by the machining process but can be made by pressure moulding in a hydraulic press (~150 tonne force at 380 °C) using a pre-machined die of appropriate size.

To make a diamond-shaped mask from metal (Fig. 2), the workpiece is held at an angle of 45° to the axis of the bed of the milling machine such that the diagonal of the mask runs parallel to the axis of the bed. The cavity is made by machining to the required depth (0.5–0.75 mm) to create a reaction chamber. The sealing edge (0.3–0.5 mm wide) is formed by machining the outlands to a depth of approximately 0.5 mm and is finished by polishing flat with successively finer grades of wetted abrasive paper from ~P600–P1200 (3M Inc., USA). The final polish is achieved by using a polishing grade crocus paper (A. J. Naylor & Co, England). Holes for reagent inlet and outlet are drilled, respectively, at the bottom and the top of the reaction chamber (in the corners of the diamond). For masks made with PTFE, the holes should be 1.0 mm diameter: for aluminium or stainless steel, the holes should be 1.07–1.08 mm diameter. Inlet and outlet connections to the synthesiser are made using standard 19SWG syringe needles (1.1 mm diameter) with the chamfered ends ground and de-burred. With these dimensions, a leak-tight seal is usually produced without the need for an additional sealer.

3.1.3
Making Scanning Arrays on an ABI DNA/RNA Synthesiser

The arrays are made on an adapted ABI DNA synthesiser using standard nucleotide CE phosphoramidites. The process of making arrays has been described in detail previously [3]. Briefly it is as follows: The mask is fixed to the frame (Fig. 2) and the inlets and outlets are connected to the reagent supply and waste lines of the DNA synthesiser using syringe needles. The mask is sealed against glass or polypropylene substrate by applying pressure (for example with a G-clamp), creating a cavity of the shape of the mask through which the reagents for DNA synthesis flow. DNA synthesis thus proceeds over a confined area on the

Table 1. Modified program for an ABI394 DNA/RNA synthesiser to deliver reagents for one coupling cycle

Step No.	Function No.	Function	Step time[a]
1	106	Begin	
2	103	Wait	999.0
3	64	18 to waste	5.0
4	42	18 to column	30.0
5	2	Reverse flush	8.0
6	1	Block flush	5.0
7	101	Phos prep	3.0
8	111	Block vent	2.0
9	58	Tet to waste	1.7
10	34	Tet to column	1.0
11	33	B+tet to column	3.0
12	34	Tet to column	1.0
13	33	B+tet to column	3.0
14	34	Tet to column	1.0
15	33	B+tet to column	3.0
16	34	Tet to column	1.0
17	103	Wait	75.0
18	64	18 to waste	5.0
19	2	Reverse flush	10.0
20	1	Block flush	5.0
21	42	18 to column	15.0
22	2	Reverse flush	8.0
23	63	15 to waste	5.0
24	41	15 to column	15.0
25	64	18 to waste	5.0
26	1	Block flush	5.0
27	103	Wait	20.0
28	2	Reverse flush	10.0
29	1	Block flush	5.0
30	64	18 to waste	5.0
31	42	18 to column	12.0
32	2	Reverse flush	8.0
33	42	18 to column	12.0
34	2	Reverse flush	8.0
35	42	18 to column	12.0
36	2	Reverse flush	8.0
37	42	18 to column	12.0
38	2	Reverse flush	8.0
39	1	Block flush	3.0
40	62	14 to waste	5.0
41	40	14 to column	35.0
42	103	Wait	20.0
43	1	Block flush	5.0
44	64	18 to waste	5.0
45	42	18 to column	25.0
46	2	Reverse flush	10.0
47	42	18 to column	25.0
48	2	Reverse flush	10.0
49	1	Block flush	3.0
50	107	End	

[a] Step times are for a diamond-shaped mask having 0.73 mm depth × 30 mm diagonal and have to be adjusted for each mask.

surface of the solid substrate. After a nucleotide has been added, the mask is displaced by a pre-determined step size, by moving the solid substrate in front of the mask in an appropriate horizontal direction. The movement is such that the back of the mask overlaps a portion of the front of its previous footprint. The mask is sealed against the substrate again for addition of the next base in the sequence. The first longest oligonucleotide in the set is made when the mask intersects its first footprint for the last time.

Routinely, we make scanning arrays covering around 150 nt of the complement of the target sequence. The process of array fabrication has now been automated; the movement of the solid substrate along the mask and the opening and closing of the mask against the substrate are operated by computer-controlled motors. This should, in principle, allow fabrication of longer arrays scanning the complement of an entire target sequence.

The cycle used to make arrays is slightly modified from that used with columns. An example of such a cycle is given in Table 1. The flow times depend on the size and the depth of the mask. It is difficult to monitor the efficiency of coupling from trityl yield due to the large amounts of reagents used in the synthetic process. However, several criteria suggest a high stepwise yield. First, the ratio of coupling reagents to the substrate is more than 200-fold greater in this process than in conventional solid-phase synthesis on a controlled pore glass (CPG) column. Secondly, the synthesis of any oligonucleotide on the array is dependent on successful coupling of each nucleotide. For example, each nucleotide in a 20-mer on the array occurs in 19 other 20-mers. Thus, successful synthesis of any one oligonucleotide (indicated by a positive hybridisation signal) will indicate successful synthesis in these others. Thirdly, synthesis of oligonucleotides using a cleavable linker and analysis by gel electrophoresis has suggested efficient yields (S. C. Case-Green and E. M. Southern, unpublished data).

It is also possible to make arrays of oligonucleotides with several other compatible chemistries. For example, with the use of standard nucleotide CE phosphoramidites in synthesis, the oligonucleotides are attached to the solid support at their 3'-end; reverse phosphoramidites (Glen Research) can be used to make oligonucleotides that are attached at their 5'-end. Such orientation may be important in applications requiring enzymatic manipulations (e. g., nucleotide addition using a DNA polymerase). Iodine is used as an oxidising agent to produce phosphodiester bonds; this can be replaced with a sulfurising agent (Beaucage reagent) to produce phosphorothioate bonds in the arrayed oligonucleotides. Modified nucleotides (e. g., 2'O-methyl-modified) can be used to obtain other variations on the standard format.

3.1.4
Deprotection of Arrays

When using standard phosphoramidite chemistry to make oligonucleotides, the exocyclic amines of the bases are protected chemically to prevent side reactions from occurring during synthesis. These protecting groups must be removed from the coupled bases before hybridisation. We achieve this by either of the following two methods: (1) by leaving the arrays in an aqueous solution of ammonia at 55 °C for 14–16 h in a closed chamber, or (2) by incubating the array in an equal volume mixture of ethanolamine and methanol at 65 °C for 30 min [4]. We have found that the first method is not suitable for deprotecting arrays made on glass derivatised using the 'vapour-phase' method (see Sect. 3.1.1).

3.2
Hybridisation of a Labeled Target to Scanning Arrays

The target nucleic acid is labeled with a radioisotope (^{33}P or ^{32}P) or a fluorescent compound (e.g., CY5) and hybridised to a scanning array. The choice of label depends on the application. For example, when studying folding of a target nucleic acid, it may be more appropriate to label it with a radioisotope rather than CY5 because the bulky CY5 groups may interfere with or alter the folding of the labeled compound.

The labeled product is generated by the standard in vitro transcription in the presence of a labeled precursor (such as [α-^{33}P]UTP or [α-^{32}P]UTP or CY5-UTP), using an appropriate DNA template. In vitro transcription is generally carried out using T7, T3 or SP6 RNA polymerase. A plasmid containing the desired fragment under the transcriptional control of a bacteriophage promoter (Table 2) can be used as a template. The plasmid is linearised with an appropriate restriction endonuclease to produce transcripts of a defined length without contaminating vector sequence. If the fragment of interest has not been cloned into such a plasmid, the template for use with a bacteriophage RNA polymerase can also be generated using a polymerase chain reaction (PCR): primers used in the PCR reaction are such that one of them has a bacteriophage promoter/leader sequence added at the 5'-end (Table 2).

Approximately 50–70 fmol of labeled transcript are used in hybridisation. The transcript is diluted in an appropriate volume of the hybridisation buffer. For arrays made on glass, the hybridisation solution is applied as a thin film of liquid between the array and a glass plate of the same size. Polypropylene arrays are hybridised in a rotating glass tube placed in a hybridisation oven at an appropriate temperature.

Arrays made on both glass and polypropylene can be stripped of target and reused several times. We have used the arrays made on polypropyene in as many as 7–8 hybridisations without significant loss of signal, after which they tend to deteriorate very quickly. Stripping solution is 100 mM sodium carbonate/bicarbonate buffer at pH 9.8–10 which is heated to ~90 °C and the arrays is dipped into it for 2–3 min. The arrays are stored in absolute ethanol at 4 °C.

Table 2. Bacteriophage promoter and leader sequences[a]

T7	5' taa tac gac tca cta ta *ggg cga*
(or)	5' taa tac gac tca cta ta *ggg aga*
T3	5'aat taa ccc tca cta aa *ggg aga*
SP6	5' att tag gtg aca cta ta *gaa tac*

[a] The hexa-nucleotide leader sequence (in *italics*) appears in the transcript.

3.3
Reading a Scanning Array Image

After hybridisation with a radiolabeled target, exposure to a storage phosphor screen captures the hybridisation signal from the array. The screen is then scanned on a PhosphorImager (such as the STORM; Molecular Dynamics). For a fluorescently labeled target, the array is scanned directly on a PhosphorImager. Visual inspection of the image reveals the results generally, but computer-assisted analysis is needed to extract quantitative information about the hybridisation capacities of the various oligonucleotides and their exact sequence and length.

We use a software package called xvseq to analyse scanning array images that is run on a SUN Solaris workstation (available by anonymous ftp at (ftp://bioch.ox.ac.uk/pub/xseq.tar.gz)). The programme reads and displays im-

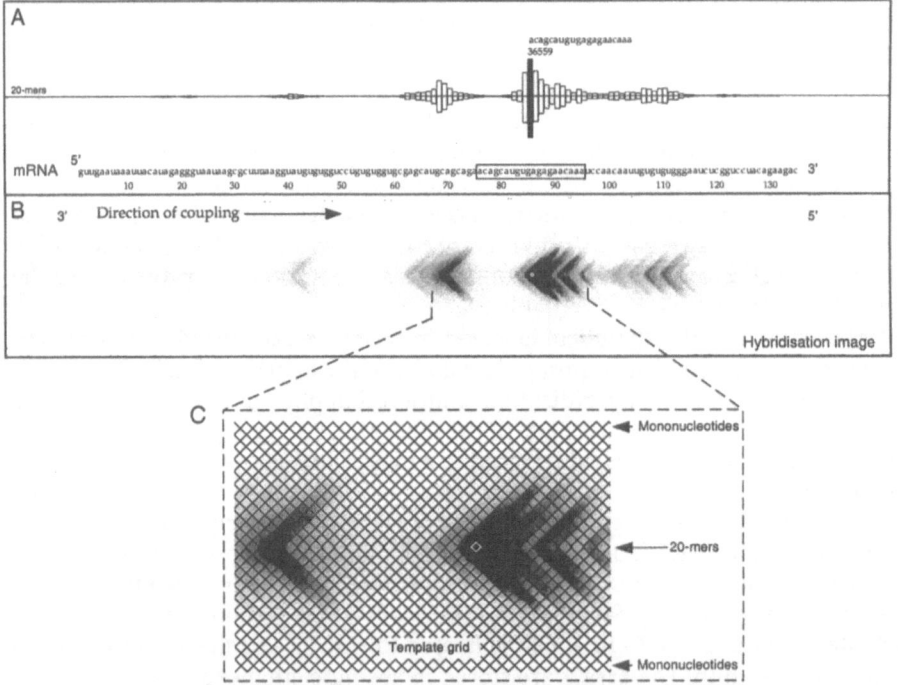

Fig. 3. Display and analysis of a scanning array image using xvseq. **A** The results of the integrated pixel values are displayed as paired histograms and the two values are for the halves above and below the centre line of the array. Only values for 20-mers are displayed here but the programme can display all values from monomers to the longest in the set. The programme also has several other useful features. For example, clicking on a cell in the array highlights the corresponding region in the sequence, the relevant bar in the histogram and the integrated pixel values of the two cells. Clicking on a bar in the histogram (*filled bar*) highlights the corresponding cells of the array and the sequence (*boxed sequence*). **B** Image of a scanning array obtained after hybridisation of a 32P-labeled transcript. **C** A part of the image magnified with a template grid overlay

ages generated by a Molecular Dynamics PhosphorImager and can perform standard image manipulations such as scaling, clipping, rotation and colormap control. Its main purpose, however, is to calculate and display integrated intensities of array oligonucleotides. The programme generates a template grid of overlapping diamonds (or circles) that is superimposed on the array image, and allows the user to specify the template size, shape (circular or diamond-shaped), location, step size between successive templates, as well as the sequence that generated the array pattern. The template grid allows accurate registration of the oligonucleotides on the array. An example of analysis of an image using this program is given in Fig. 3.

4
Applications

4.1
High-Throughput Screening of Antisense Reagents

Antisense reagents comprise two major classes, antisense oligonucleotides and ribozymes. Antisense oligonucleotiodes act by binding to the target mRNA and mediating its cleavage at the heteroduplexed site by an RNAse H-based mechanism, or by strongly binding to the target site and thus interfering with its translation. Ribozymes are catalytic RNA molecules that are capable of mediating cutting at the target sequence. Such reagents can therefore be used, in principle, to explore function of a gene, and are potential therapeutics for many diseases caused by abnormal or overexpression of genes, such as those involved in cancers or those of viral origin. Both types of antisense reagents require a site on the target mRNAs where they can bind efficiently. Finding such sites is not an easy task; it is generally found that only 5% – 10% of the chosen sites allow efficient binding by the antisense reagents and, in addition, antisense reagents placed only a single or a few bases apart can have quite different antisense activities. There is now convincing evidence that intramolecular folding in RNAs renders most of the molecule inaccessible to intermolecular binding [5 – 10]. However, folding of RNAs is difficult to predict [11] and the rules that govern heteroduplex formation are poorly understood [12]. Effective antisense reagents are thus found empirically, and a large number of reagents need to be individually synthesised and screened for their binding capacities with the target sequence. A full review of the screening methods to find effective antisense reagents is beyond the scope of this article and is presented elsewhere (e. g., [13 – 15]).

The scanning arrays provide an efficient method to screen for open structures in RNAs that can be targeted with antisense reagents and allow a large number of reagents to be screened simultaneously in simple hybridisation tests. This method provides quantitative measurements of the relative accessibility of every site in a RNA and thus facilitates selection of the optimal target sequence for designing antisense reagents. Hybridisation to scanning arrays is typically carried out in 1 M NaCl, which is far from being physiological. However, we have found an excellent correlation between the binding shown by oligonucleotides on arrays under these conditions and their antisense activities in many systems [8, 16,

17). Milner et al. [8] used β-globin mRNA as target and showed that duplex yield on a scanning array of anti-β-globin antisense oligonucleotides correlated with their inhibitory potential in an in vitro translation assay. We have also tested the antisense activities of over 30 oligonucleotides targeted to mRNAs of cyclins B1, B4 and B5 of *Xenopus laevis* in cell-free egg extracts and oocytes [17]. Scanning arrays corresponding to the first 120 nucleotides of each of the three B-type cyclins were hybridised to their corresponding transcript. Only some of the oligonucleotides were found to produce strong heterduplexes and in many cases a single or a few bases shift produced significant changes in the heteroduplex yields. Oligonucleotides that produced strong heteroduplex yield, and others that showed little annealing, were assayed for their effect on translation of endogenous cyclin mRNAs in Xenopus egg extracts and their ability to promote cleavage of cyclin mRNAs in oocytes by RNase H. Excellent correlation was found between the antisense potency and the affinities of the oligonucleotides for cyclin transcripts as measured by the arrays, despite the complexity of the cellular environment. Although a role for cellular proteins and other small molecules in influencing the binding of antisense reagents to their targets has been proposed [18, 19], surprisingly our results suggest that it may not be a major factor.

4.2
The Study of Nucleic Acid Folding and Heteroduplex Formation

In order to develop theoretical methods to predict intermolecular binding, one needs to understand not only the folding process of nucleic acids, but also the mechanism of duplex formation. Scanning arrays have proved to be useful tools to study both these aspects.

4.2.1
RNA Folding

Folding pathways of large RNAs (as large as mRNAs) are poorly understood. There are two generally held views: one is that most secondary interactions are local and are established as RNA is transcribed; the alternative view is that RNAs (including mRNAs) exist in a global minimum free energy state – a state that would have to be established after the synthesis was complete. However, there is little evidence to support either view. We used a scanning array to explore these suggestions. With compelling evidence that intermolecular binding largely depends on folding of the interacting species [5 – 10], we assumed that the interaction of short oligonucleotides with an RNA could be used as a tool to study changes in the folding of an RNA molecule [20]; we speculated that a change in the secondary structure of an RNA would also result in alteration of its interaction with complementary oligonucleotides – accessible regions may become inaccessible, and vice versa.

To mimic the synthetic pathway, we made transcripts that had a common start point (with one exception) but widely different end points by transcribing restriction fragments with different lengths; the other had a small deletion at the 5'-end. These transcripts were hybridised to an array that had oligonucleotide complementary to a part of the common sequence shared by the transcripts. We expected that, if these transcripts were to reach a global free energy minimum, the transcripts with different sizes would acquire different secondary structures and so produce different hybridisation patterns. Theoretical free energy calculations of the transcripts did predict quite different secondary structures. Strikingly, the predominant features of the hybridisation pattern were shared by all the transcripts. These data suggested that RNA folding creates local stable states that are trapped early in the transcription or folding process and that these states are not altered to produce structures with global minimum free energy.

4.2.2
Heteroduplex Formation

For two main reasons it is difficult to predict the sites on a RNA that would hybridise with antisense reagents. First, despite the elegance of the computational methods, there is a great deal of uncertainty in the prediction of folding in mRNAs. Second, the rules that govern intermolecular binding are also poorly understood. If predictive approaches are to be useful, both these aspects need to be fully understood.

Mir and Southern [12] studied the mechanisms of heteroduplex formation by studying the binding of oligonucleotides to structurally well-characterised tRNA^{phe} in solution. In an exhaustive assay they used an array of 12-mer (and shorter) oligonucleotides complementary to tRNA^{phe} to measure binding to radiolabeled tRNA^{phe}. The hybridisation data were analysed on the basis of the clover-leaf secondary and three-dimensional crystal structures of tRNA^{phe}. They derived certain rules from this study. They found that most of the tRNA was inaccessible and as small as a single base shift produced significant changes in the hybridisation yields. All regions of tRNA that formed heteroduplexes had unpaired bases stacked onto the end of a stem and most of the bases that were incorporated into the heteroduplex were part of a continuous helix within the tRNA and on one side of the stem. Strikingly, there were several other oligonucleotides with similar features and overlapping sequences that did not produce heteroduplexes with tRNA. The oligonucleotides that gave the highest heteroduplex yields were accommodated within the existing higher-order stacking architecture of the native structure. They also showed that binding of oligonucleotides could be affected by oligonucleotide length as well as sequence. Overall, the factors that influenced the binding were embedded in the structure of tRNA.

4.2.3
A Study of the Secondary Structure of Nucleic Acids with Modified Bases

Nguyen and Southern [21] used scanning arrays to study the hybridisation of a target DNA containing modified bases. The aim of this investigation was to find ways to destabilise the secondary structure in the target that has such a strong influence on intermolecular binding. As a model, they used a DNA target with a self-complementary sequence forming a stem loop. They replaced dC with N-4-ethyldeoxycytidine (d^{4Et}C) that specifically hybridises with natural dG to give a

G:^{4Et}C base pair but with reduced stability compared with the normal G:C base pair. By hybridising it to an array of complementary oligonucleotides, they found that this substitution greatly weakened the target's secondary structure and allowed it to hybridise to the complementary oligonucleotides more strongly than with the natural analogue.

5
Concluding Remarks

Scanning arrays are versatile tools for the analysis of nucleic acid interactions. These can therefore provide a platform to improve the array-based methods. The applications of scanning arrays discussed in this article are only those that have been published to date. Undoubtedly, new applications of this method will emerge in the future and combination of scanning arrays with conventional approaches may provide a more powerful way forward in the study of nucleic acid interactions.

Acknowledgements. We would like to thank all past and present members of the group who contributed to the development of scanning arrays and Martin Johnson for his description of mask construction. The Medical Research Council, UK, funded M. Sohail.

6
References

1. Maskos U, Southern EM (1992) Nucleic Acids Res 20:1679
2. Matson RS, Rampal JB, Coassin PJ (1994) Anal Biochem 217:306
3. Sohail M, Southern EM (2001) In: Rampal JB (ed) Methods in molecular biology: oligonucleotide arrays – methods and protocols, vol 170. Humana Press, pp 181–199
4. Polushin NN, Morocho AM, Chen B-C, Cohen JS (1994) Nucleic Acids Res 22:639
5. Lima WF, Monia BP, Ecker DJ, Frier SM (1992) Biochemistry 31:12055
6. Southern EM, Case-Green SC, Elder JK, Johnson M, Mir KU, Wang L, Williams JC (1994) Nucleic Acids Res 22:1368
7. Bruice TW, Lima WF (1997) Biochemistry 36:5004
8. Milner N, Mir KU, Southern EM (1997) Nat Biotechnol 15:537
9. Patzel V, Sczakiel G (1998) Nat Biotechnol 16:64
10. Sohail M, Akhtar S, Southern EM (1999) RNA 5:646
11. Michel F, Westhof E (1990) J Mol Biol 216:585
12. Mir KU, Southern EM (1999) Nat Biotechnol 17:788
13. Branch A (1998) Antisense Nucleic Acid Drug Dev 8:249
14. Sohail M, Southern EM (2000) Curr Opin Mol Ther 3:264
15. Sohail M, Southern EM (2000) Adv Drug Deliver Rev 44:23
16. Southern EM, Milner N, Mir KU (1997) In: Chadwick DJ, Cardew G (eds) Ciba Foundation Symposium 209, John Wiley, Chichester, p 38
17. Sohail M, Hochegger H, Klotzbucher A, Guellec RL, Hunt T, Southern EM (2001) Nucleic Acids Res 29:2041
18. Bertrand EL, Rossi JJ (1994) EMBO J 13:2904
19. Herschlag D, Khosla M, Tsuchihashi Z, Karpel RL (1994) EMBO J 13:2913
20. Zarrinkar PP, Williamson JR (1994) Science 265:918
21. Nguyen H-K, Southern EM (2000) Nucleic Acids Res 28:3904

Received: June 2001

The Use of MassARRAY Technology for High Throughput Genotyping

Christian Jurinke · Dirk van den Boom · Charles R. Cantor · Hubert Köster

SEQUENOM, Inc., 3595 John Hopkins Court, San Diego, CA 92121,USA.
E-mail: cjurinke@sequenom.com

This chapter will explore the role of mass spectrometry (MS) as a detection method for genotyping applications and will illustrate how MS evolved from an expert-user-technology to a routine laboratory method in biological sciences. The main focus will be time-of-flight (TOF) based devices and their use for analyzing single-nucleotide-polymorphisms (SNPs, pronounced snips). The first section will describe the evolution of the use of MS in the field of bioanalytical sciences and the protocols used during the early days of bioanalytical MALDI TOF mass spectrometry. The second section will provide an overview on intraspecies sequence diversity and the nature and importance of SNPs for the genomic sciences. This is followed by an exploration of the special and advantageous features of mass spectrometry as the key technology in modern bioanalytical sciences in the third chapter. Finally, the fourth section will describe the MassARRAY technology as an advanced system for automated high-throughput analysis of SNPs.

Keywords: SNP, High-throughput, MALDI mass spectrometry, Genomics, DNA chip

1	**Mass Spectrometry in the Biological Sciences**	58
1.1	Mass Spectrometry	58
1.2	ESI MS	58
1.3	MALDI-TOF MS	59
1.4	Mass Spectrometry in the Protein World	61
1.5	Mass Spectrometry in the Carbohydrate World	61
1.6	Mass Spectrometry in the DNA World	62
2	**Polymorphic Markers in the Genomic Sciences**	63
2.1	Intraspecies Sequence Variations	63
2.2	Restriction Fragment Length Polymorphisms and Short Tandem Repeats	64
2.3	Single Nucleotide Polymorphisms	64
2.4	Pharmacogenomics	65
3	**The Advantages of Mass Spectrometry for Genetic Diversity Detection**	66

Advances in Biochemical Engineering/
Biotechnology, Vol. 77
Managing Editor: T. Scheper
© Springer-Verlag Berlin Heidelberg 2002

4   MassARRAY and High-Throughput Genotyping 67

5   Summary and Future Perspectives 72

6   References 72

List of Symbols and Abbreviations

ESI   Electrospray ionization
LCR   Ligase chain reaction
MALDI   Matrix assisted laser desorption/ionization
MS   Mass spectrometry
PCR   Polymerase chain reaction
PSD   Post source decay
RFLP   Restriction fragment length polymorphism
SNP   Single nucleotide Polymorphism
STR   Short tandem repeat
TOF   Time-of-flight
VNTR   Variable number tandem repeats

1
Mass Spectrometry in the Biological Sciences

1.1
Mass Spectrometry

The general principle of mass spectrometry (MS) is to produce, separate and detect gas phase ions. Traditionally, thermal vaporization methods are used to
transfer molecules into the gas phase. The classical methods for ionization are
electron impact (EI) and chemical ionization (CI). Most biomolecules, however,
undergo severe decomposition and fragmentation under the conditions of both
methods. Consequently, the capabilities of mass spectrometry have been limited
to molecules the size of dinucleotides [1]. Analysis of oligonucleotides with a
mass range of up to 3000 Da became feasible with the development of plasma
desorption (PD) methods [2]. However, until the invention of 'soft' ionization
techniques such as ESI- and MALDI MS, mass spectrometric tools were not
widely considered for routine applications in biological sciences.

1.2
ESI MS

Electrospray ionization (ESI) was first introduced by Fenn et al. in the late 1980s
[3]. The method is based on spraying an analyte through a syringe into an electrical field. Ionization of analyte molecules consists of coulomb explosions, resulting in fully desolvated ions migrating towards the mass analyzer. The spectra obtained by ESI MS are mostly composed of multiple charged ions of the

parent molecules. The analyte mass is determined by deconvolution of the complex spectra. Application of ESI MS to the analysis of nucleic acids has been described by several authors [4, 5]. However, today the most common bioanalytical applications of ESI MS are in the field of peptide and protein analysis [6, 7]. The main reasons for this may be the higher duty cycle of MALDI MS, an important factor for routine high-throughput applications, and the complex signal patterns obtained from ESI MS combined with a comparably low tolerance for impurities.

1.3
MALDI-TOF MS

Matrix assisted laser desorption/ionization (MALDI) as a principle for analysis of large biomolecules was introduced by Karas and Hillenkamp in 1989 [8]. Briefly, in MALDI MS, the sample is embedded in the crystalline structure of small organic compounds (matrix) and deposited on a conductive sample support. The co-crystals are irradiated with a nano-second laser beam, for example a UV laser with a wavelength of 266 nm or 337 nm. The energies introduced are in the range of 1×10^7 to 5×10^7 W/cm^2. The laser energy causes structural decomposition of the irradiated crystal and generates a particle cloud (the plume) from which ions are extracted by an electric field. The mechanism behind this not fully understood process of desorption may be best described as a conversion of laser energy to vibrational oscillation of the crystal molecules which finally results in the disintegration of the crystal. After acceleration, the ions drift through a field-free path (usually one to two meters in length) and finally reach the detector (e.g. a secondary electron multiplier or channel plate) (see Fig. 1). Ion masses (mass-to-charge ratios, m/z) are typically calculated by measuring their time-of-flight (TOF) which is longer for larger molecules than for smaller ones (provided their initial energies are identical). Since single charged, non-fragmented ions are predominantly generated, parent ion masses can easily be determined from the resulting spectrum without the need for complex data processing. The masses are accessible as numerical data for direct processing and subsequent analysis. TOFs measured during a typical MALDI experiment are in the range of several microseconds.

The quality of the spectra, which is reflected in terms of resolution, mass accuracy, signal-to-noise ratio, and also sensitivity, is highly dependent on sample preparation and on the choice of matrix compounds [9]. The predominant effects influenced by sample purification and matrix choice are adduct formation and fragmentation. Both will be described in detail in the following sections. In brief, the purification process should result in a sample which is free of crystallization disturbing agents (like detergents, urea or DMSO), adduct forming agents (such as unvolatile cations) and not too diluted. The classical method of preparing a sample on a holder, still widely used today, is the so called 'dried-droplet' method [10]. It relies simply on pipetting a small volume of the sample (usually about 0.5 µl) to a drop (about 1 µl) of matrix solution and allowing the mixture to dry. For high throughput applications, as described later, this procedure has to be modified.

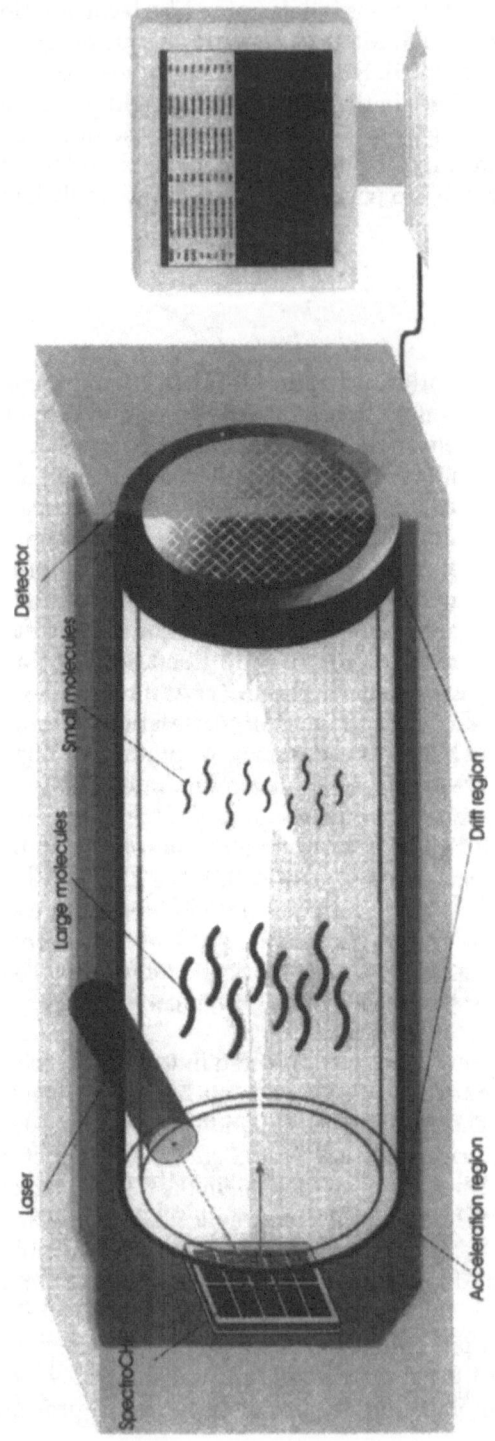

Fig. 1. Schematic representation of MALDI-TOF mass spectrometer

1.4
Mass Spectrometry in the Protein World

Both, ESI and MALDI TOF MS have been proven as valuable research tools for the analysis of peptides and proteins and also for the investigation of their interactions with each other or different substances. The characterization of proteins was one of the first described applications of MALDI MS. Proteins and peptides are easier to volatilize than nucleic acids. Besides analyzing peptide patterns from cell lysates or serum, recent interesting applications include the analysis of patterns from differentiated tissues of organs and the typing of bacteria using whole cells. An exciting demonstration of the analytical power of MALDI MS is the investigation of the neurotransmitter content of single neurons [11].

An advantage of MALDI compared to ESI is that MALDI can deal with significantly higher amounts of impurities. For example, it is possible to analyze hemoglobin proteins directly from blood [12]. The analysis of milk proteins without any purification has also been demonstrated [13]. Caprioli and coworkers described a direct profiling of proteins in tissue sections [14]. The authors were able to distinguish between proximal, intermediate, and distal colon from fresh tissue blotted on a conductive polyethylene membrane. Prior to this work, Eckerskorn et al. demonstrated that MALDI analysis of blotted proteins after gel electrophoresis is feasible [15]. Protein- and peptide conjugates are another field of application where MS techniques provided valuable contributions. An example is the analysis of antibody conjugates by MALDI TOF [16]. Furthermore, an emerging field is the analysis of non-covalent protein-DNA complexes and interactions [17, 18].

1.5
Mass Spectrometry in the Carbohydrate World

Another area of bioanalytics where MALDI TOF plays an important role is composition analysis of complex carbohydrates or glyco-conjugates such as glycopetides or glycoproteins. A very basic application is the analysis of complex carbohydrate solutions [19]. A more sophisticated application is the compositional analysis of carbohydrate components of complex proteins, for example elucidation of glycosylation sites [20] and composition analysis of carbohydrate side chains. For this purpose, enzymatic methods are combined with a special feature of MALDI, post source decay [21]. As described in Sect. 1.6, fragmentation of analyte molecules can occur throughout the MALDI process, even if the desorption and ionization is comparably soft. If this fragmentation occurs after the ions have left the source area of the mass spectrometer, it is referred to as post-source-decay (PSD). Special instrumentation (ion gates) are used to allow only ions of defined flight-time to proceed towards the detector. With this method, fragments deriving from a certain parent ion can be analyzed. Reflectron instruments have to be employed to resolve and detect the ion subspecies. Using this instrumentation, oligosaccharide mixtures cleaved from glycopeptides by treatment with enzymes like PNGase can be analyzed. By careful interpretation (using spectra libraries for comparison) the structural composition of the glycan side chains can be revealed [22].

1.6
Mass Spectrometry in the DNA World

The application of MALDI MS to the analysis of nucleic acids has for years remained a field of small impact for the bio-science community. Nucleic acids are far more difficult to analyze under MALDI conditions than peptides, so applications have been limited. The reason is that nucleic acids are especially susceptible to adduct formation and fragmentation. Because of their negatively charged phosphate backbone, nucleic acids are polyanions. As such, they tend to form salt adducts with cations present in the surrounding medium. In biochemical reactions, these are predominantly sodium and potassium ions. The result of adduct formation is a distribution of the signal: a main signal with the protonated analyte may be accompanied by signals resulting from multiple adduct formation. For example, every sodium ion attached to the analyte molecule will cause an additional signal with a mass of plus 23 Da. Consequently, adduct formation lowers sensitivity and analytical accuracy – the total amount of ions is distributed over a multitude of ion species, and the mass differences between those may become too small to be resolved. Approaches to overcome adduct formation are purification and ion-exchange procedures based either on solid-phase [23], or cation exchange resin methods [24]. The chemical modification of the phosphate backbone has also been proposed for preventing adduct formation [25]. The addition of ammonium containing additives such as di-ammonium citrate [26] or tartrate [27] to the matrix has a huge impact on cation heterogeneity of analytes. The use of mM solutions of ammonium hydroxide during the conditioning process has also a beneficial effect [28]. The reason for exchanging cations for ammonium is that the latter is a volatile cation which is released as ammonia in the gas phase leaving the analyte molecules as free acids.

Based on the chemical nature of nucleic acids, some fragmentation reactions can occur relatively efficiently during the MALDI process. The predominant fragmentation is depurination.

Protonization of the nucleobases A and G (at position N7) induces a polarization of the N-glycosidic bond between sugar and nucleobase finally resulting in nucleobase elimination. Subsequently to depurination further fragmentation can occur via backbone cleavage. Also the loss of C has been reported [29] which is mediated through protonation at N3. It has been demonstrated by Hillenkamp and coworkers that RNA is comparably more stable under MALDI conditions than DNA. The proposed reason is the lack of the 2′ hydroxy group in the ribose sugar moiety [30].

Approaches to compensate for the detrimental influence of adduct formation have been contributed by instrument developers and molecular biologists. It has been demonstrated by several groups that introduction of 7-deaza nucleobases, where N7 is replaced by C, is helpful in suppressing depurination reactions [31, 32], since the proton acceptor site is removed.

An important contribution to analyzing DNA with MALDI-TOF is also the introduction of delayed extraction instruments by Wiley and McLaren [33]. Compared to static extraction, which employs a permanent electrical field for ion acceleration, the delayed extraction accumulates ions over a time range of some nanoseconds before the extraction voltage is applied. The effect is a compensation of the initial differences in kinetic energies of the analyte molecules, since molecules with higher initial velocities expand further during the delay time and experience a lower potential during acceleration. To some extent, this may also compensate for positional variations of molecules which occur due to an inhomogeneous distribution in the matrix cocrystals.

The introduction of favorable matrixes [34] for DNA analysis has led to the development of many different applications ranging from oligonucleotide sequence analysis [35, 36] and PCR- or LCR product detection [37, 38] to PCR product sequencing [39], mutation detection [40, 41] and clinical diagnostics [42, 43]. Furthermore, strategies to analyze combined amplification and sequencing reactions have been described [44] and the analysis of in vitro transcripts has also been reported [45]. On the more analytical side, fragmentation mechanisms for nucleic acids under MALDI conditions [46] and the conditions for the analysis of intact double-stranded DNA have been investigated [47]. Some of the latest and most promising lines of development are the application of IR-MALDI for the analysis of DNA molecules of up to 2.1 kb by Hillenkamp and coworkers [48] and the description of bench-top MALDI instruments for biopolymer analysis by Cotter and coworkers [49].

2
Polymorphic Markers in the Genomic Sciences

2.1
Intraspecies Sequence Variations

Within the foreseeable future, the efforts of the Human Genome Project (HGP) will lead to a reference sequence for *Homo sapiens*. A 'working version' of the human genome sequence – covering about 97% of the entire genome – was presented to the public on the 26 of June 2000, jointly from scientists of the HGP and Celera. Today, the research organizations associated in the HGP produce about 1000 bases of raw sequence data per second – 24 hours a day and seven days a week. With this dedicated effort, we can expect the remaining gaps to be closed by 2003 or earlier. However, our knowledge of the architecture and function of the human genome is poor – there are still disputes on whether there are 40,000 or 140,000 genes in the human genome [50] and on the role of the so called 'junk DNA'. Another blind spot on our map is the encoding of intraspecies genetic diversity. We estimate that the genetic diversity within the human species is about 0.1 to 0.3%. For purpose of comparison, the genetic identity between *Homo sapiens* and both *Pan* species (Chimpanzees and Bonobos) is about 98.4% – which is closer than the relationship between African and Indian elephants.

The knowledge of the 3.2 billion base pair human sequence information and information on our genetic diversity are milestones in our understanding of the contribution of genes to many complex problems such as disease predisposition and diversity of drug response by different patients. The manifestations of genetic diversity within the human genome are mostly SNPs, RFLPs (which can be treated somehow as a subset of SNPs), STRs and random mutations. The nature and importance of these polymorphic genetic elements are discussed in the following sections.

2.2
Restriction Fragment Length Polymorphisms and Short Tandem Repeats

RFLPs (restriction fragment length polymorphisms) can be defined as (single) nucleotide polymorphisms located in the recognition or cleavage site of a restriction endonuclease, thus altering the corresponding cleavage pattern. With this definition, RFLPs are bi-allelic markers with the allele status permissive to cleavage, yes or no. Although RFLP markers [51] are useful in many applications, they are often of poor information content, for example in family linkage studies and their analysis is cumbersome to automate.

STR markers (short tandem repeats, microsatellites) [52], in contrast, are quite highly informative (through their polymorphic number of repeats) and easy to prepare and analyze using PCR-based assays with a considerable potential for automation. However, with conventional gel electrophoresis-based analysis, typing of large numbers of individuals for hundreds of STR markers still remains a challenging task. Today, the application of STR analysis is mainly in the field of identification (forensics) and agricultural breeding approaches, covering both, plants and animals.

The development of markers to characterize intraspecies sequence variations has, from a more theoretical point of view, evolved from the use of RFLPs to microsatellites and very recently to SNPs.

2.3
Single Nucleotide Polymorphisms

Single nucleotide polymorphisms (SNPs) can be defined as bi-allelic base variants within a population occurring with an allelic frequency higher than 1% (with 1% as a more or less arbitrary threshold). SNPs can be seen as a generalization on single nucleotide polymorphisms, covering also RFLPs.

The importance of SNPs to our understanding of genetic diversity is recognized today by many academic and commercial organizations. Consequently this is leading to huge SNP discovery efforts within the human genome. The program which has received the most public attention is The SNP Consortium (TSC), a collaboration between 13 multinational companies and prominent academic institutions like the Washington University, the Whitehead Institute and The Sanger Centre. The declared goal of TSC is to deliver 300,000 SNPs to the public domain by the end of 2001. Currently, more than 100,000 SNP have been disclosed to the public [53].

The Human Genome Project (HGP) also has an ongoing human diversity project, which will lead to a genome-wide map of 100,000 SNPs [54].

The strategies for discovering SNPs are either direct or indirect approaches. For a direct approach, certain regions of the genome, e.g. genes which may be interesting disease predisposition candidates, are chosen and screened thoroughly for all SNPs occurring in many different individuals. This kind of study is referred to as a high saturation study and will lead to a defined set of SNPs. Since the target genes are pre-selected, there is a high probability of identifying a disease-causing SNP by selecting those polymorphisms which alter amino acid (coding- or cSNPs) or regulatory sequences for subsequent association studies.

Another, more indirect and broad approach is based on mining polymorphisms from existing data bases by comparing overlapping sequences. This approach results in so called *in silico* SNPs (isSNPs). The isSNPs need to be validated or confirmed since they can result either from sequencing errors or real polymorphisms. For the validation of isSNPS, real DNA samples need to be analyzed.

The technologies used to discover SNPs are diverse and include re-sequencing (either gel- or microchip hybridization based), SSCP, HPLC and enzymatic cleavage methods [55, 56].

In whatever manner they have been discovered, SNPs need to be analyzed among many different populations to reveal their meaning for human healthcare. A direct way to benefit from information about human diversity is to compare the allelic frequencies of SNPs in populations of individuals, either affected or unaffected by a certain disease. Candidate genes, supposed to be causative for the disease of interest are carefully selected by analyzing metabolic pathways which are involved in the etiology of the particular disease. Such a study results in the association of certain SNPs with the manifestation of a clinical status. This approach as for example resulted in knowledge about susceptibility to HIV infection, which has been demonstrated [57] to correlate to the genetic status of a chemokine receptor (CCR5). Another example is the correlation of deep-vein thrombosis to a polymorphic variation in coagulation factor gene FV [58].

A more indirect way to capitalize from knowledge about human diversity is to analyze the reasons for differences in drug metabolism; this approach is called pharmacogenomics.

2.4
Pharmacogenomics

Traits within populations, such as the ABO blood groups, are phenotypic expressions of genetic diversity. This is also the case for certain variations in response to drug therapy. When taken by poor metabolizers, some drugs may cause exaggerated pharmacological response and adverse drug reactions. For example, tricyclic antidepressants exhibit order-of-magnitude differences in blood concentrations depending on the status of patients' metabolizing enzymes [59]. Pharmacogenomics and pharmacogenetics are often used as synonyms. However, pharmacogenetics is the study of genetic polymorphism in drug metabolism and pharmacogenomics is more an umbrella term for the optimization of drug discovery and development using genetic knowledge. The major idea behind the efforts of such development is to provide people with individualized drugs. Instead of having one block buster drug which should fit for all patients and ignoring adverse drug reactions, pharmaceutical companies are now thinking about providing drugs especially for certain populations. For example, pharmaceutical companies can screen individuals for specific genetic polymorphisms before entry into clinical trials to ensure that the study population is both relevant and representative. Targets for such screenings are cytochrome P450 enzymes or *N*-acetyltransferase isoenzymes (NAT1 and 2). Potential drug candidates affected by polymorphic metabolism include antidepressants, antipsychotics and cardiovascular drugs. Another possible application deriving from knowledge of genetic diversity is the potential of rescuing drugs for the market. A promising drug candidate, which fails to get approval from regulatory authorities because of certain side-effects may enter the market with a restricted approval for application on patients with a well-defined genetic background. At first glance, this may sound like a softening of the restrictive approval procedures. However, by taking into account that these drugs will only be prescribed after careful examination of the individual patient, this concept opens the way to bring highly efficient drugs to people, with the knowledge that unwanted side effects will be excluded. Since the money spent on the development of failed drugs finally is amortized through approved drugs, a positive side effect of this strategy may be that drug treatment, over all, will become cheaper.

Of course, the methods employed for such approaches need to fulfill strict quality requirements, since the outcome of an error is potentially fatal.

3
The Advantages of Mass Spectrometry for Genetic Diversity Detection

The requirements for SNP discovery- and SNP typing (or scoring) technologies are quite different. All of the above-mentioned methods for SNP discovery have merits. For large scale SNP scoring projects, however, they will fail to be efficient. The reasons are a lack of reasonable automation and an inherent susceptibility to errors. The first issue is mainly due to gel electrophoresis technology which is cumbersome to automate. Another issue is caused by indirect detection methods based on labels.

The inherent and clearly predominant advantage of mass spectrometry as a detection device is a direct detection of the analyte itself. This is a paradigm shift for molecular biology, since it eliminates all uncertainties caused by indirect detection via labels. This may be illustrated by an example. Consider a certain PCR amplicon which is subject to unbalanced PCR (one allele is amplified with significant preference). Consider further, that mismatches can occur if hybridization is used to identify the alleles of a SNP. This generates some background signal. With these two inherent processes in mind, one can easily imagine that there is a point where the signal intensities generated by either effect can no longer be distinguished. That means a true heterozygote affected by imbalanced PCR generates the same signal as a mismatched probe. This effect needs to be compensated by a redundant experiment setup and careful statistical analysis. From pure analytical and also practical points of view these are drawbacks which can be overcome, if MS is applied. The dynamic range of analyzing MassEXTEND reactions with MALDI TOF allows for the detection of allele ratios of up to 1:50. The analytical accuracy of MALDI TOF is about 0.1 to 0.01% of the determined mass. This means a 15mer DNA with about 4500 Da can unambiguously be distinguished from a 16mer which is approximately 300 Da, or 6.6%, higher in mass.

A major factor which needs to be taken into account is the cost of a whole genotyping project – not just the cost of individual SNP typing reactions. A major factor in these costs of course is data analysis by highly trained specialists. Furthermore, every new SNP assay needs a considerable time for implementation using standard curves and statistics. These are all certain drawbacks in a high throughput environment. Assay design for MassEXTEND reactions can be done automatically and implementation steps as necessary for fluorescence-based assays are obsolete.

An important feature in the cost per assay is the capability of easy multiplexing. Multiplexed assays are conducted together in one tube (or one well of a micro plate) and combined; for example, assays are combined for different SNPs to save time and reagents. With fluorescent detection, the level of multiplexing is somewhat coupled to the amount of available dyes – which is in most applications only four. For MALDI-based SNP detection, a multiplex level of up to 12 assays has already been described [60].

These advantages taken together may be the reason that mass spectrometric based methods have been referred to as the 'gold standard' for SNP analysis in a recent review on this topic [61].

4
MassARRAY and High-Throughput Genotyping

The MassEXTEND reaction was developed by Sequenom especially for the purpose of assessing genetic polymorphism by mass spectrometry. The assay format can be used for the analysis of deletion-, insertion-, and point-mutations, STR-, and SNP-analysis, and it allows for the detection of compound heterozygotes. The MassEXTEND process comprises a post-PCR primer extension reaction carried out in the presence of one or more dideoxynucleotides (ddNTPs) and generates allele-specific terminated extension fragments (see Fig. 2). For SNP analysis the primer binding site is placed adjacent to the polymorphic position. Depending on the nucleotide status of the SNP, a shorter or a longer extension product is generated. In the case of heterozygosity, both products are generated. After completion of the reaction, the products are purified and analyzed by MALDI TOF MS [62].

In the example given in Fig. 2, both elongation products are expected to differ in mass by one nucleotide. Raw data for a heterozygous DNA sample analyzed by a MassEXTEND assay are given in Fig. 3. The two SNP alleles appear as two distinct mass signals. Careful assay design makes a high-level multiplexing of MassEXTEND reactions possible.

In the case of STR analysis, a ddNTP composition is chosen which terminates the polymerase extension at the first nucleotide not present within the repeat [63]. For length determination of a CA repeat a ddG or ddT termination mix is used. Even imperfect repeats harboring insertion or deletion mutations can be analyzed with this approach. Figure 4 displays raw data from the analysis of a human STR marker in a heterozygous DNA sample. Both alleles differ by 4 CA repeats. The DNA polymerase slippage during amplification generates a pattern of 'stutter fragments' (marked with in asterisk in Fig. 4). In case of heterozygotes

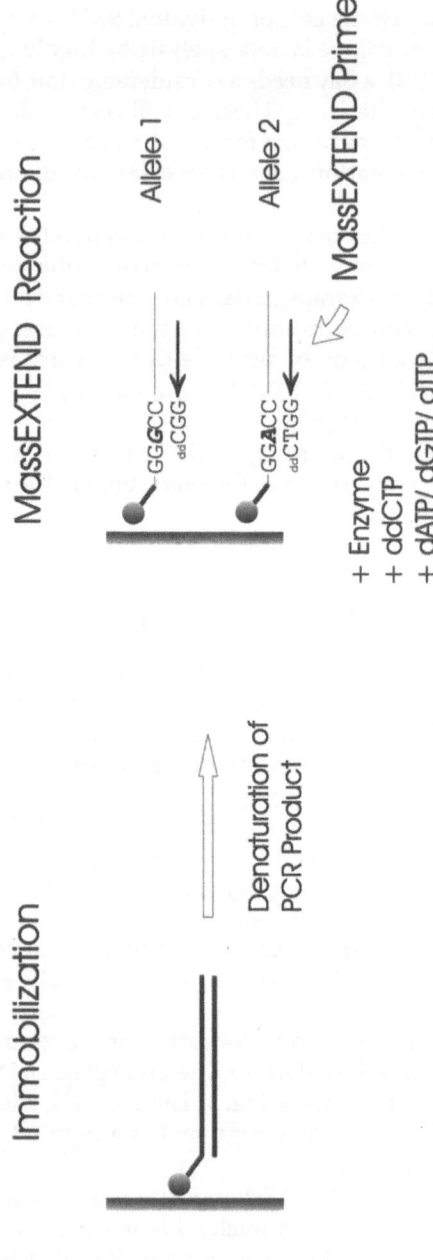

Fig. 2. Reaction scheme for MassEXTEND reaction

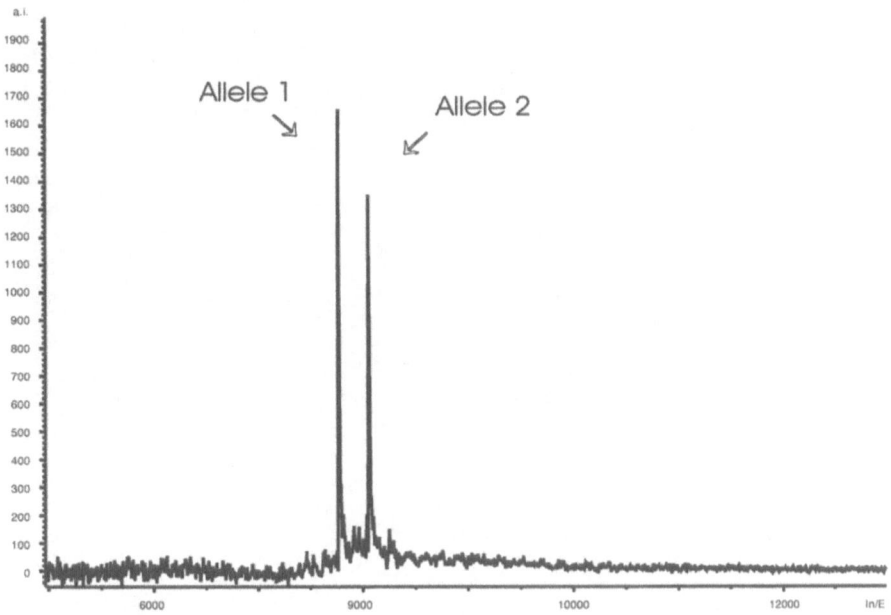

Fig. 3. Raw data for a typical MassEXTEND reaction

which differ in just one repeat, the smaller allele has higher intensities than the larger allele, because allelic and stutter signals are added together.

When compared to the analysis of hybridization events by detecting labels – even on arrays, the DNA MassARRAY approach differs significantly. The Mass-EXTEND assay is designed to give only the relevant information. The mass spectrometric approach enables a direct analyte detection with 100% specificity and needs no redundancy. This accuracy and efficacy is combined with sample miniaturization, bioinformatics and chip-based technologies for parallel processing of numerous samples .

The use of an advanced nano-liquid handling combined with surface modified silicon chips permit today the automated scanning of 384 samples in about

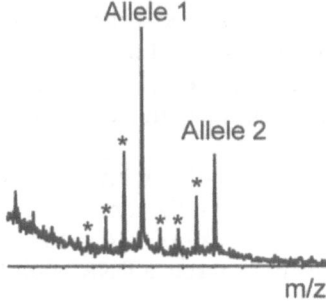

Fig 4. Raw data of MassEXTEND reaction for STR analysis

Fig. 5. SpectroCHIP with 384 different samples

40 minutes. From the mass spectrometric point, the advantage of this sample preparation method is the generation of extremely homogeneous crystals allowing for an automated scanning with just ten laser shots per sample. Currently up to 10 SpectroCHIPs each containing up to 384 different samples (see Fig. 5) can be analyzed in one unattended run using a Bruker/SEQUENOM SpectroREAD mass spectrometer. The SpectroREAD addresses each position of the chip sequentially, collects the sum of ten laser shots, processes and stores the data and proceeds to the next spot of the chip.

Using a proprietary algorithm, masses as well as signal-to-noise ratios are automatically analyzed and interpreted. After completion of analysis, the results are transferred to a data base and stored as accessible genetic information (see Fig. 6). The database also provides a tool for visual control and comparison of spectra with theoretically expected results. MassEXTEND assays can easily be designed in a high throughput mode with a computer aided assay development module.

A very promising tool in combination with MassARRAY is the analysis of allelic frequencies by means of pooled DNA samples. In this approach a significant number of different genomic DNA samples (e.g. 100 individual DNAs) are combined in equimolar amounts to generate a DNA pool. This pooled population is subjected to PCR amplification and analyzed in a single MassEXTEND reaction. If carefully analyzed, the outcome of such an experiment is information about the

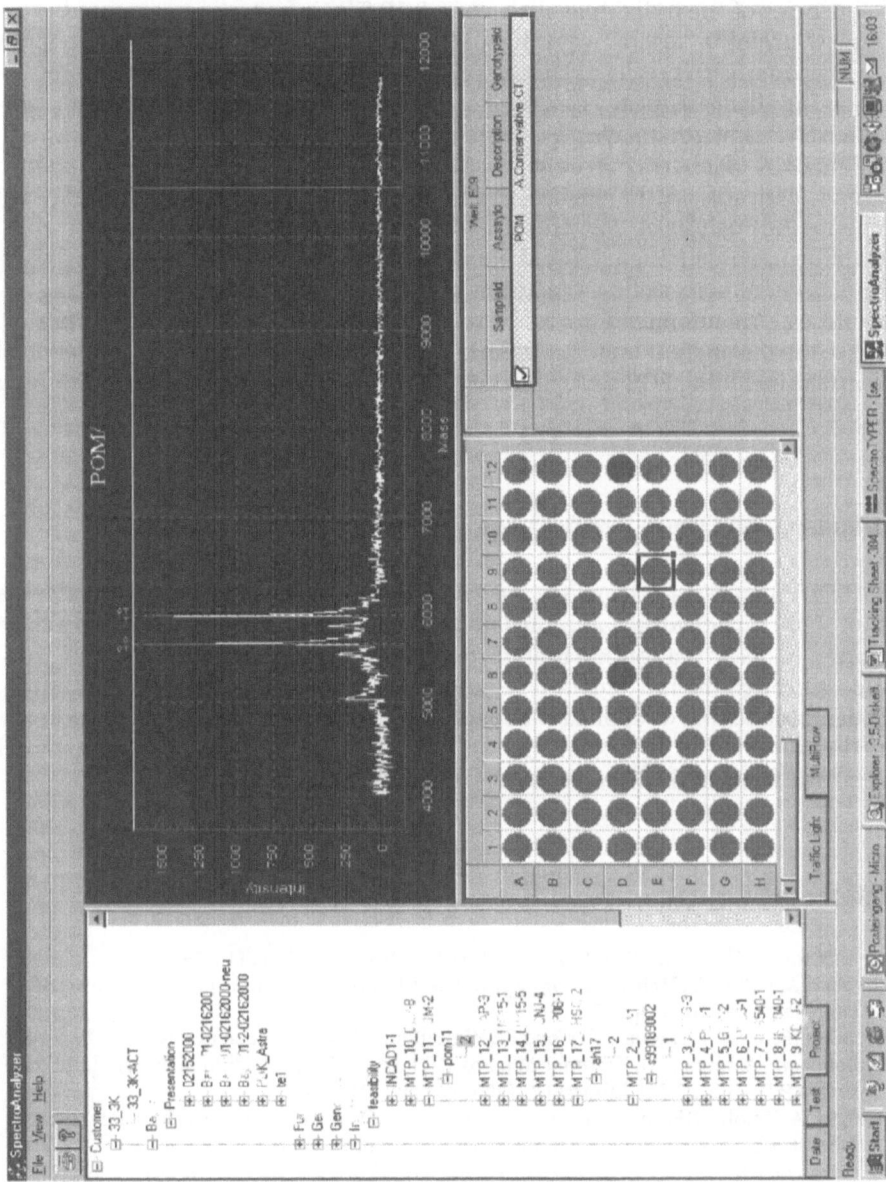

Fig. 6. Screenshot of processed mass spectrum with genotype analysis

frequencies of the individual alleles of a SNP. The information generated is extremely useful in several ways. It can be used to validate a SNP – to decipher whether an isSNP is really a relevant polymorphism rather than a sequencing artifact. The allelic frequency can, furthermore, be used as a filter to decide if a candidate SNP is worth analysis in larger extend populations.

The DNA MassArray throughput, measured in terms of genetic information output, depends on the chosen scale. Using microtiter plates and 8-channel pipettes the analysis of 192 genotypes (two 96 well MTPs) a day is routine work. With the use of automated liquid handling stations and 384 MTPs the throughput has been increased by a factor of about 30 (not including multiplexing capabilities). An automated process line (APL) has been developed to further increase the throughput up to an industrial scale. The APL integrates biochemical reactions including PCR set-up, MassEXTEND reaction and sample conditioning into a fully automated process.

5
Summary and Future Perspectives

MALDI TOF mass spectrometry is a valuable tool for modern bioanalytical tasks in protein, carbohydrate, and nucleic acid analysis. MS-based techniques have evolved to become a routine high throughput device for analyzing genetic variations through the introduction of automated chip-mediated sample scanning. With a broadening of applications especially in the field of DNA analysis and a further automation in protein analysis MALDI MS is on track to become the gel electrophoresis of the 21st century.

6
References

1. McCloskey JA, Crain PF (1992) Int. J. Mass Spectrom. Ion Proc. 118:593
2. Viari A, Ballini JP, Meleard P, Vigny P, Dousset P, Blonski C, Shire D (1988) Biomed. Environ. Mass Spectrom. 16:225
3. Fenn JB, Mann M, Meng CK, Wong SF, Whitehouse CM (1989) Science 246:64
4. Johnson YA, Nagpal M, Krahmer MT, Fox KF, Fox A (2000) J. Microbiol. Methods 40:241
5. Null AP, Hannis JC, Muddiman DC (2000) Analyst 125:619
6. Loo JA, DeJohn DE, Du P, Stevenson TI, Ogorzalek Loo RR (1999) Med. Res. Rev. 19:307
7. Veenstra TD (1999) Biochem. Biophys. Res. Commun. 257:1
8. Karas M, Hillenkamp F (1988) Anal. Chem. 60:2299
9. Börnsen KO, Gass MAS, Bruin GJM, von Adrichem JHM, Biro MC, Kresbach GM, Ehrat M (1997) Rapid Commun. Mass Spectrom. 11:603
10. Nordhoff E, Ingendoh A, Cramer R, Overberg A, Stahl B, Karas M, Hillenkamp F, Crain PF (1992) Rapid Commun. Mass Spectrom. 6:771
11. Garden RW, Moroz LL, Moroz TP, Shippy SA, Sweedler JV (1996) J. Mass Spectrom. 31:1126
12. Houston CT, Reilly JP (1997) Rapid Commun. Mass Spectrom. 13:1435
13. Siciliano R, Rega B, Amoresano A, Pucci P (2000) Anal. Chem. 72:408
14. Chaurand P, Stoeckli M, Caprioli RM (1999) Anal. Chem. 71:5263
15. Eckerskorn C, Strupat K, Karas M, Hillenkamp F, Lottspeich F (1992) Electrophoresis 13:664

16. Siegel MM, Tabei K, Kunz A, Hollander IJ, Hamann RR, Bell DH, Berkenkamp S, Hillenkamp F (1997) Anal. Chem. 69:2716
17. Deterding LJ, Kast J, Przybylski M, Tomer KB (2000) Bioconjug. Chem. 11:335
18. Yates JR 3rd (2000) Trends Gent. 16:5
19. Charlwood J, Tolson D, Dwek M, Camilleri P (1999) Anal. Biochem. 273:261
20. Korsmeyer KK, Guan S, Yang ZC, Falick AM, Ziegler DM, Cashman JR (1998) Chem. Res. Toxicol. 11:1145
21. Spengler B (1997) J. Mass Spectrom. 32:1019
22. Rouse JC, Strang AM, Yu W, Vath JE (1998) Anal. Biochem. 256:33
23. Tang K, Fu DJ, Kötter S; Cotter RJ, Cantor CR, Köster H (1995) Nucleic Acids Res. 23:3126
24. Nordhoff E, Ingendoh A, Cramer R, Overberg A, Stahl B, Karas M, Hillenkamp F, Crain PF (1992) Rapid Commun. Mass Spectrom. 6:771
25. Gut IG, Beck S, (1995) Nucleic Acids Res. 23:1367
26. Pieles U, Zürcher W, Schär M, Moser HW (1993) Nucleic Acids Res. 21:3191
27. Lecchi P, Pannell LK (1995) J. Am. Soc. Mass Spectrom. 6:972
28. Jurinke C, van den Boom D, Collazo V, Lüchow A, Jacob A, Köster H (1997) Anal. Chem. 69:904
29. Gross J, Strupat K (1998) Trends Anal. Chem. 17:470
30. Nordhoff E, Cramer R, Karas M, Hillenkamp F, Kirpekar F, Kristiansen K, Roepstorff P (1993) Nucleic Acids Res. 21:3347
31. Kirpekar F, Nordhoff E, Kristiansen K, Roepstorff P, Hahner S, Hillenkamp F (1995) Rapid Commun. Mass Spectrom. 9:525
32. Siegert C, Jacob A, Köster H (1996) Anal. Biochem. 243:55
33. Wiley WC, McLaren IH (1953) Rev. Sci. Instrum. 26:1150
34. Wu KJ, Steding A, Becker CH (1993) Rapid Commun. Mass Spectrom. 7:142
35. Pieles U, Zurcher W, Schar M, Moser HE (1993) Nucleic Acids Res. 21:3191
36. Faulstich K, Worner K, Brill H, Engels JW (1997) Anal. Chem. 69:4349
37. Tang K, Taranenko NI, Allmann SL, Ch'ang LY, Chen CH (1994) Rapid Commun. Mass Spectrom. 8:727
38. Jurinke C, van den Boom D, Jacob A, Tang K, Wörl R, Köster H (1996) Anal. Biochem. 237:174
39. Koster H, Tang K, Fu DJ, Braun A, van den Boom D, Smith CL, Cotter RJ, Cantor CR (1996) Nat. Biotechnol. 14:1123
40. Little DP, Braun A, Darnhofer-Demar B, Frilling A, Li Y, McIver RT Jr, Koster H (1997) J. Mol. Med. 75:745
41. Higgins GS, Little DP, Koster H (1997) Biotechniques 23:710
42. Jurinke C, Zöllner B, Feucht HH, van den Boom D, Jacob A, Polywka S, Laufs R, Köster H (1998) Genet. Anal. 14:97
43. Braun A, Little DP, Köster H (1997) Clin. Chem. 43:1151
44. van den Boom D, Ruppert A, Jurinke C, Köster H (1997) J. Biochem. Biophys. Meth. 35:69
45. Kirpekar F, Nordhoff E, Kristiansen K, Roepstorff P, Lezius A, Hahner S, Karas M, Hillenkamp F (1994) Nucleic Acids Res. 22:3866
46. Gross J, Leisner A, Hillenkamp F, Hahner S, Karas M, Schafer J, Lutzenkirchen F, Nordhoff E (1998) J. Am. Soc. Mass Spectrom. 9:866
47. Little DP, Jacob A, Becker T, Braun A, Darnhofer-Demar B, Jurinke C, van den Boom D, Köster H (1997) Int. J. Mass Spectrom. Ion Processes 169:133
48. Berkenkamp S, Kirpekar F, Hillenkamp F (1998) Science 281:260
49. Fancher CA, Woods AS, Cotter RJ (2000) J. Mass Spectrom. 35:157
50. Aparicio SAJR, (2000) Nat. Genet. 25:127
51. de Martinville B, Wyman AR, White R, Francke U (1982) Am. J. Hum. Genet. 34:216
52. Taylor GR, Noble JS, Hall JL, Stewart AD, Mueller RF (1989) Lancet 2:454
53. http://snp.cshl.org/
54. Collins F S, Patrinos A, Jordan E, Chakravarti A, Gesteland R, Walters L, and the members of DOE and NIH planning groups (1998) Science 282:682

55. Lindblad-Toh K, Winchester E, Daly MJ, Wang DG, Hirschhorn J, Laviolette JP, Ardlie K, Reich DE, Robinson E, Sklar P, Shah N, Thomas D, Fan JB, Gingeras T, Warrington J, Patil N, Hudson TJ, Lander ES (2000) Nat Genet 24:381
56. Underhill PA, Jin L, Lin AA, Mehdi SQ, Jenkins T, Vollrath D, Da RW, Cavalli-Sforza LL, Oefner PJ (1997) Genome Res 7:996
57. Gonzalez E, Bamshad M, Sato N, Mummidi S, Dhanda R, Catano Cabrera S, McBride M, Cao XH, Merrill G, O´Connell P, Bowden DW, Freedman BI, Anderson SA, Walter EA, Evans JS, Stephan Clark RA, Tyagi S, Ahuja SS, Dolan MJ, Ahuja SK (1999) Proc Natl Acad Sci USA 96:12004
58. Bertina RM (1997) Clin Chem 43:1678
59. Larrey D (1989) Hepatology 10:168
60. Ross P, Hall L, Smirnov I, Haff L (1998) Nat. Biotechnol. 16:1347
61. Weaver T (2000) Trends Genet. In press
62. Braun A, Little D, Köster H (1997) Clin Chem 43:1151
63. Braun A, Little D, Reuter D, Müller-Mhysok B, Köster H (1997) Genomics 46:18

Received: June 2001

Sequencing by Hybridization (SBH): Advantages, Achievements, and Opportunities

Radoje Drmanac · Snezana Drmanac · Gloria Chui · Robert Diaz · Aaron Hou · Hui Jin · Paul Jin · Sunhee Kwon · Scott Lacy · Bill Moeur · Jay Shafto · Don Swanson · Tatjana Ukrainczyk · Chongjun Xu · Deane Little

Callida Genomics, 670 Almanor Ave., Sunnyvale, CA 94085, USA. *E-mail: rade@sbh.com*

Efficient DNA sequencing of the genomes of individual species and organisms is a critical task for the advancement of biological sciences, medicine and agriculture. Advances in modern sequencing methods are needed to meet the challenge of sequencing such megabase to gigabase quantities of DNA. Two possible strategies for DNA sequencing exist: direct methods, in which each base position in the DNA chain is determined individually (e.g., gel sequencing or pyrosequencing), and indirect methods, in which the DNA sequence is assembled based on experimental determination of oligonucleotide content of the DNA chain. One promising indirect method is sequencing by hybridization (SBH), in which sets of oligonucleotides are hybridized under conditions that allow detection of complementary sequences in the target nucleic acid. The unprecedented sequence search parallelism of the SBH method has allowed development of high-throughput, low-cost, miniaturized sequencing processes on arrays of DNA samples or probes. Newly developed SBH methods use DNA ligation to combine relatively small sets of short probes to score potentially tens of millions of longer oligonucleotide sequences in a target DNA. Such combinatorial approaches allow analysis of DNA samples of up to several kilobases (several times longer than allowed by current direct methods) for a variety of DNA sequence analysis applications, including de novo sequencing, resequencing, mutation/SNP discovery and genotyping, and expression monitoring. Future advances in biochemistry and implementation of detection methods that allow single-molecule sensitivity may provide the necessary miniaturization, specificity, and multiplexing efficiency to allow routine whole genome analysis in a single solution-based hybridization experiment.

Keywords: Sequencing, Hybridization, Oligonucleotide, Arrays, Ligation

1	**Introduction**	76
1.1	Direct and Indirect DNA Sequencing	76
1.2	History of Sequencing by Hybridization	78
2	**Principles of SBH**	79
2.1	Proper Probe Length for SBH	79
2.2	Oligonucleotide Overlap Principle	79
2.3	Branching Points and Multiple Assembly Solutions	81
2.4	Universal and Target-Specific Sets of Probes	82
2.5	Probe Design and Hybridization Specificity	83
2.6	Highly Parallel Data Collection in Different Formats	84

Advances in Biochemical Engineering/
Biotechnology, Vol. 77
Managing Editor: T. Scheper
© Springer-Verlag Berlin Heidelberg 2002

3   **SBH Studies on Arrays of Samples** 85

3.1   SBH on DNA Arrays 85
3.2   Blind De Novo Sequencing Test on Arrays of Samples 85
3.3   Sample Arrays for Novel Polymorphism and Mutation Discovery 86

4   **SBH on Arrays of Oligonucleotides** 87

4.1   Initial SBH Tests on Probe Arrays 87
4.2   Use of Non-Universal Arrays of Probes 88
4.3   Hybridization Combined with Polymerase 89
4.4   Hybridization Combined with Ligase: Combinatorial Probe Scoring 89
4.5   A Combinatorial SBH Blind Test 90
4.6   De Novo Sequencing, Comparative Sequencing and Genotyping
         Using Combinatorial Ligation and the HyChip System 91

5   **Future Research Directions** 95

5.1   Hybridization Technologies 95
5.2   Hybridization Chemistry 97
5.3   Advanced SBH Applications 98

6   **Conclusion** 99

7   **References** 99

1
Introduction

1.1
Direct and Indirect DNA Sequencing

Accurate DNA sequencing is a crucial procedure in modern biological, medical and agricultural research. Traditionally, DNA sequencing has been done by direct, base-by-base analysis, with each new base determination built on the results of many previous sequencing steps (Fig. 1). Such direct methods are effective but result in a gradual accumulation of errors in the sequencing process. Extreme experimental precision may help minimize these errors, but generally direct sequencing methods are restricted to relatively short (usually 300–1000 bases) DNA targets to maintain accuracy. This article discusses sequencing by hybridization (SBH), an indirect sequencing approach that avoids many of the shortcomings of the direct methods.

Direct sequencing techniques involve a variety of synthesis, degradation, or separation techniques, and include the traditional Sanger [1], pyrosequencing [2, 3] and exonuclease methods [4], as well as direct visualization approaches [5–8]. In the Sanger method, DNA synthesis is randomly terminated at each base pair, creating a wide range of fragments that are then separated by gel according to length and scored. In the pyrosequencing method, polymerase-guided incorpo-

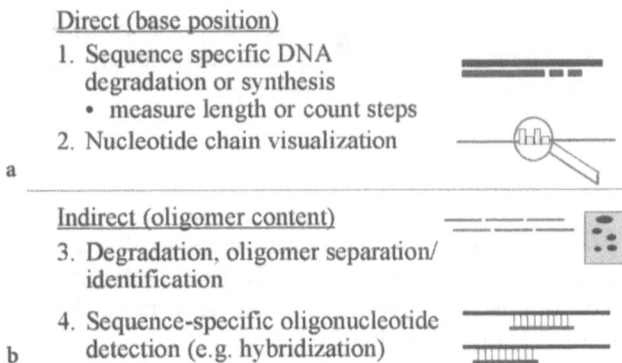

Direct (base position)
1. Sequence specific DNA
   degradation or synthesis
   • measure length or count steps
2. Nucleotide chain visualization

a

Indirect (oligomer content)
3. Degradation, oligomer separation/
   identification

4. Sequence-specific oligonucleotide
b   detection (e.g. hybridization)

Fig. 1. DNA sequencing strategies. Sequencing methods are either direct or indirect, depending on the type of experimental data obtained from a DNA target. **a** Direct methods involve determination of which character or base is present in a particular position in the DNA string. There are two major types of direct sequencing. The first involves sequence-specific DNA degradation, synthesis, or separation followed by base-by-base assignment; the second involves direct chain visualization. **b** Indirect sequencing methods involve determination of which substrings of bases are present or absent in the DNA target. The first of two indirect approaches involves fragmentation of target DNA, followed by separation and identification of the resulting oligonucleotides. Sequencing by hybridization (SBH), the subject of this article, is the second indirect method. It involves sequence-specific pairing of DNA targets with complementary oligonucleotide probes of known or determinable sequence

ration of each base is detected by measuring the pyrophosphate released in consecutive cycles. In the exonuclease approach, a single molecule of target DNA synthesized with fluorescent-tagged bases is degraded by exonuclease. The consecutively released nucleotides are then scored with a very sensitive fluorescence detector. An interesting new class of direct sequencing methods relies on base visualization, using either scanning-tunneling microscopy [5–7] or nano-pores [8]. The main technical difficulties associated with direct base visualization are incorrect base differentiation and low speed.

Indirect sequencing methods such as SBH have several unique advantages and have become important tools in deriving accurate DNA sequence information. Indirect sequencing methods are based on determining the oligonucleotide content of a target nucleic acid and do not require determination of base position information experimentally (Fig. 1). Oligonucleotide content may be obtained in two principal yet unrelated ways, by fragmenting the DNA and separating the resulting oligonucleotides based on physiochemical properties, or by specific hybridization of oligonucleotides to complementary sequences present in the test DNA. The fragmentation method was actually the first nucleic acid sequencing method [9–11], developed before the currently used Sanger gel-based method. Sequencing by hybridization was first proposed in 1987 [12, reviewed in 13]. In SBH methods, numerous oligonucleotides are tested for their ability to hybridize to a target DNA. Each oligonucleotide that successfully hybridizes to the target indicates the presence of a complementary DNA sequence within the target, revealing a small piece of information about its sequence. The complete sequence is then determined by compiling the results of many such hybridization tests. The SBH method has the potential to accurately and inexpensively sequence very long nucleic acids (>10,000 bases) in a single reaction.

1.2
History of Sequencing by Hybridization

Scientists from many disciplines helped construct the intellectual framework that led to the development of SBH in 1987 [12]. In 1960, Doty observed that, when heated in solution, double-stranded DNA "melts" to form single stranded chains, which then re-nature spontaneously when the solution is cooled [14]. This result suggested the possibility of using one piece of DNA to recognize another. In the late 1970s, researchers discovered ways to efficiently synthesize large quantities of short, sequence-specific DNA molecules by stepwise addition of nucleotides to columns [15]. At the same time, Bruce Wallace used oligonucleotides as "hybridization probes" to confirm the existence of complementary sequences in target DNAs, and demonstrated that they can accurately detect single base mutations in defined DNA targets dot-blotted on membranes [16]. In 1986, Lehrach proposed using a small set of oligonucleotide probes for cosmid clone fingerprinting [17]. At about the same time, PCR was developed by Mullis [18]. These early experiments provided basic research tools that were later fundamental in developing methods for fingerprinting and sequencing DNA by oligonucleotide hybridization.

Sequencing by hybridization was first proposed and patented by Drmanac and Crkvenjakov in 1987 [12]. They proposed that the sequence of a DNA fragment could be determined by a hybridization experiment in which the target DNA was exposed to a set of oligonucleotides under conditions that favored full match hybridization. The sequences of the subset of positively hybridizing probes would then be used to determine the target sequence. One format proposed by the researchers was hybridization of labeled probes or various probe pools to large arrays of clones as a method for sequencing complex DNA samples. They presented an algorithm for reconstruction of DNA sequences as complex as the human genome [19]. As with the PCR method, SBH dramatically expanded potential applications of the DNA hybridization reaction.

Additional research studies followed in the SBH area. Bains and Smith demonstrated sequence reconstruction in a simulated assembly process using 256 specially designed 6-mers with two degenerate base positions in the middle (gapped probes) [20]. In a 1988 patent application, Southern proposed a method for combinatorial in situ synthesis of complex oligonucleotide arrays on glass [21]. He proposed that DNA samples could be hybridized to the probe array for mutation detection and complete sequencing. Lysov and colleagues also proposed DNA sequencing by hybridization to probe arrays, in this case using arrays of oligonucleotides deposited on a solid support [22]. Macevicz proposed using a small probe set having a binary probe design involving a specific base (i.e. adenine) and a degenerate "non-A base" (i.e., a mix of T, C, and G) [23].

2
Principles of SBH

2.1
Proper Probe Length for SBH

When an oligomer of length n hybridizes specifically with a target DNA it reveals the existence of one or more complementary n-mer sequences within the chain. A particular n-mer sequence occurs roughly once in every $4^n/2$ base pairs of double-stranded DNA. This statistical ratio helps define appropriate design parameters for SBH experiments [19]. Long DNA molecules require longer probes, to avoid situations in which every probe has one or more complementary sequences in the target, leading to complete loss of specific information. Likewise, short probes (< 5 bases) bind so frequently to most targets that they provide little or no useful information. Very long probes, on the other hand, occur so rarely that without prior sequence information to guide probe design their use is too inefficient for most standard SBH procedures. Very large sets of long probes may be used, but this approach becomes increasingly difficult as tens of thousands of individual probes must be synthesized and arrayed. Such large probe sets may require special methods of parallel synthesis and hybridization (see Sect. 5.1, "Hybridization Technologies"). In general, SBH experiments may use probe sets of length 5 to 25 bases, though some applications may use longer probes. SBH methods can accommodate nucleic acid targets of vastly different lengths, ranging from a few bases to hundreds of megabases, depending on the experimental design [13]. Probe pools ranging in size from a few to tens of thousands of probes can also be used to minimize the number of hybridization experiments required, a method that is especially effective for shorter target nucleic acids [12, 20, 23 – 26].

2.2
Oligonucleotide Overlap Principle

In an SBH experiment involving a complete set of n-mer probes (i.e., all possible probes of length n), each DNA base is redundantly read by n overlapping probes. Figure 2 illustrates how probe overlap can be used to assemble the sequence of the target DNA in an SBH experiment using 8-mer probes. In this example, each positive probe overlaps seven bases of its nearest neighbor, six bases of its second neighbor, etc. This overlap principle allows determination of sequences that are much longer than the length of each probe by comparing and aligning $n-1$ or fewer overlapping bases that are shared by the probes. It is important to note that a full $n-1$ overlap between probes is not required to determine nucleic acid sequences; shorter overlaps may suffice, allowing for the use of incomplete probe sets. An additional advantage of probe overlap is that the effect of random errors is minimized because each base is "read" by multiple probes. Potential ambiguities in sequence resulting from the existence of long tandem repeats (e.g., $(CA)_{30}$) and $n-1$ or longer repeat regions can be re-

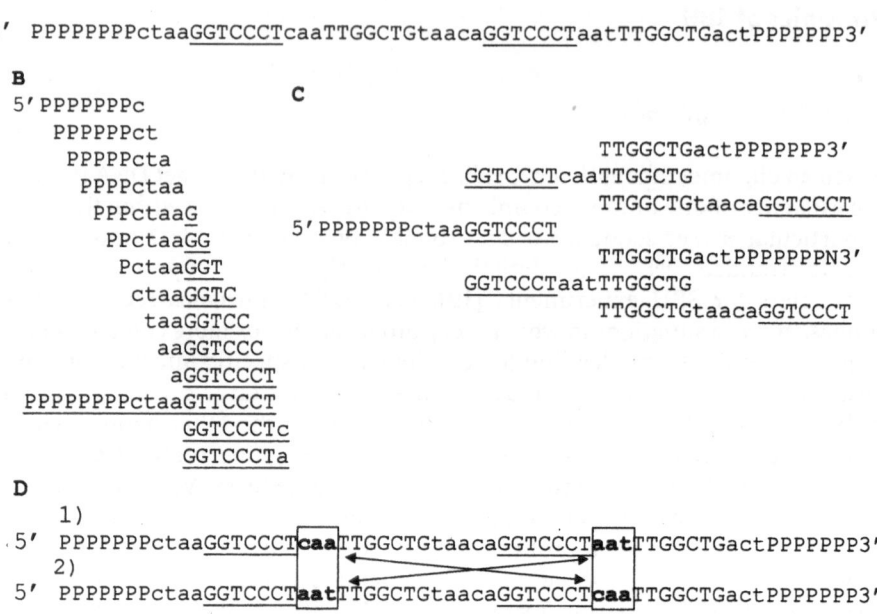

A

5′ PPPPPPPctaa<u>GGTCCCT</u>caaTTGGCTGtaaca<u>GGTCCCT</u>aatTTGGCTGactPPPPPPP3′

B
5′ PPPPPPPc
  PPPPPPct
    PPPPPcta
     PPPPctaa
      PPPctaa<u>G</u>
       PPctaa<u>GG</u>
        Pctaa<u>GGT</u>
         ctaa<u>GGTC</u>
          taa<u>GGTCC</u>
           aa<u>GGTCCC</u>
            a<u>GGTCCCT</u>
<u>PPPPPPPctaaGTTCCCT</u>
             <u>GGTCCCT</u>c
             <u>GGTCCCT</u>a

C

                              TTGGCTGactPPPPPPP3′
                      <u>GGTCCCT</u>caaTTGGCTG
                              TTGGCTGtaaca<u>GGTCCCT</u>
5′ PPPPPPPctaa<u>GGTCCCT</u>
                              TTGGCTGactPPPPPPPN3′
                      <u>GGTCCCT</u>aatTTGGCTG
                              TTGGCTGtaaca<u>GGTCCCT</u>

D
  1)
.5′ PPPPPPPctaa<u>GGTCCCT</u>**caa**TTGGCTGtaaca<u>GGTCCCT</u>**aat**TTGGCTGactPPPPPPP3′
  2)
5′ PPPPPPPctaa<u>GGTCCCT</u>**aat**TTGGCTGtaaca<u>GGTCCCT</u>**caa**TTGGCTGactPPPPPPP3′

Fig. 2. Probe overlaps and assembly branches. **A** A sample sequence containing known primer sequences PPPPPPP at each end and two 7-mer repeat sequences is shown. **B** Assembly using overlapped probes is initiated at the 5′ end by first selecting a positive probe that corresponds to the primer sequence. Sequence extension continues in the 3′ direction until the first branching point, where two possible options for further extension occur (GGTCCCTc and GGTCC-CTa). The branching process (**C**) leads to many alternative sequences. Requiring that the sequence must end with the 3′ primer and that sequence sub-fragments between branching points occur only once results in assembly of two possible sequences (**D**), which differ in placement of CAA and AAT sub-sequences

solved by additional experimental treatments such as the use of longer oligonucleotides, etc.

The process of sequence assembly from the set of oligonucleotide sequences detected in the test sample is a unique feature of SBH and other indirect methods, and allows sequencing of DNA targets that have been randomly fragmented into pieces only slightly longer than probes. It also allows researchers to sequence mixtures of fragments from different DNA molecules, provided that the total length of all segments is not too large for the given set of probes. This characteristic is an important advantage, allowing diagnostic sequencing of diverse gene segments scattered across the genome that have been amplified in a single multiplex PCR reaction.

2.3
Branching Points and Multiple Assembly Solutions

Determining a long DNA sequence from a large set of probes may pose unique technical challenges. One of these is the issue of "branching ambiguities", which may occur when three or more probes that share a common end sequence hybridize to independent sites within the target DNA. For example the probes GATAC, ATACA and ATACG all share the common four-base sequence "ATAC". These three probes represent a "branch point", a region where sequence determination can proceed in two possible directions. Without additional sequence information it is impossible to determine whether the sequence at this branch point continues as GATACA or GATACG. Fortunately the only information missing in such a situation is the correct order of otherwise correctly assembled sequence sub-fragments between branching points. This information can be obtained using various experimental or computational DNA mapping techniques. Figure 2 shows a common example of a branching ambiguity that results in two equally plausible sequences [19].

The frequency of branching points depends on probe length, target length, and bias in base content of the target DNA. For random sequences, unique assembly is expected over approximately 90% of the time for DNA targets 200, 400, 800, 1,600 and 3,200 bp in length using probes 8, 9, 10, 11, and 12 bases long, respectively [19]. Longer probes appear proportionally less efficient than short probes, because each base added results in four times as many probes but only doubles the effective DNA sequence length. This paradox results from the test requirement that a single long sequence solution must be produced. Much longer DNA or RNA samples can be analyzed if instead the goal is only to assemble a target sequence in the form of a set of sub-fragments of a given average length. For example, a set of 12-mer probes may be able to sequence over 15 kb of DNA (instead of only 3.2 kb) if 30 sequence sub-fragments of length 500 bp (a typical gel read length) are assembled instead of a single continuous 15 kb DNA molecule.

It is important to distinguish the actual number of probes in a complete probe set used in an SBH experiment from the actual number of independent probe hybridization measurements that are needed in the experiment. The number of hybridizations in an SBH experiment can be significantly reduced by using carefully designed probe pools. For example, a set of 16 million 12-mers is typically required for de novo sequencing of an unknown 3 kb piece of DNA. Fortunately a set of 16,384 oligomer pools (each containing over one thousand 12-mer probes) can be used effectively instead. Effective pool size can be increased until approximately one pool in five contains a positive probe for the DNA sample tested [25].

Long probe sets are not the only method of resolving branching ambiguities in long DNA targets. Many different experimental or data processing approaches have been suggested, including competitive hybridization [12], overlapping clones [12, 19], contiguous stacking hybridization [27], and simultaneous sequencing of similar genomes [28]. The use of closely related reference sequences (usually > 90% identity) is also useful for eliminating branching point ambiguities. Such applications include sequencing of genes from related species, such as human and primate DNA or highly conserved ribosome genes from different bacterial species. Additional procedures in which reference DNA sequences are used to guide analysis of hybridization results include complete re-sequencing of targeted human genes from different individuals, complete re-sequencing of HIV and other pathogen genes and genomes, or SNP discovery experiments focused on detection of single-base changes. SNP discovery requires the greatest accuracy since rare polymorphisms or mutations (those occurring at less than 3% frequency) are usually present in the population as heterozygotes. In addition to using reference sequences to guide sequence assembly of the test sample, probe hybridization data obtained on one or more reference or control samples may be compared with test sample data to increase the confidence of the new base calls [12, 29 – 31].

2.4
Universal and Target-Specific Sets of Probes

Universal sets of probes are designed using simple combinatorial and statistical principles, without guidance from prior knowledge of any specific gene or DNA sequence. Such probe sets may contain all possible probes of a given length, or a smaller subset of these, chosen without bias toward any specific gene sequence. An example of an incomplete universal probe set might be the set of all non-complementary probes (50% of all possible probes) used to analyze samples of denatured double stranded DNA. Another option might be a minimal set that covers each base in any sequence with at least one probe. This can be accomplished using a subset of about 3000 of the 16,000 possible 7-mer probes (unpublished data). Partial sequencing, in which only certain segments of the target DNA are determined by overlapping probes, requires even fewer probes.

Universal probe sets may be used to sequence DNA from any organism, gene or mutant, and are suitable for many different applications, including de novo and comparative sequencing, genotyping, and diagnostics. The probe sets selected can be optimized to the particular research task. Complete probe sets provide the most accurate analysis, because every target sequence, known or unknown, is covered by a maximal number of overlapping sequences.

Non-universal probe sets are chosen from larger universal sets on the basis of known sequence information from a DNA molecule of interest. In creating non-universal probe sets a variety of selection rules may be used. One common option is to use all (or a subset) of overlapping probes of a given length that are fully complementary to a known DNA sequence [31, 32]. Such sets can be further expanded by including probes that differ by one or a few bases from the known sequence [30, 32]. Because only a specific small fraction of the millions of possible probes is required, non-universal probes may be longer than those used in universal sets. This approach allows researchers to detect the presence of a known sequence in a sample, determine presence/absence and position of mutations, or discover novel SNPs or single base mutations. The main disadvantage of non-universal probe sets is the need to prepare a new set for every target.

2.5
Probe Design and Hybridization Specificity

During the past three decades, numerous experiments have demonstrated the specificity of probe hybridization [33–35]. Reproducible hybridization data, with a high average ratio of full match to mismatch scores at every base position, facilitates sequence determination using SBH. Obtaining high discrimination for long probes is more difficult, especially for those with mismatched bases at the ends of the probe. Among all eight possible mismatches G/T and G/A are the most difficult to discriminate, especially when they occur in certain 3-mer sequences (CGC/GTG and GGC/CAG) [36, 37].

In most SBH experiments, probe sets are deliberately designed to improve discrimination. The use of probes shorter than 25 bases inherently enhances discrimination. In addition, SBH experiments may use probe molecules with an informative oligomer core of length n surrounded by a few degenerate (variable) bases, e.g. $N_x - B_n - N_y$, which helps enhance discrimination [12, 19, 31]. Using degenerate bases to surround a fixed oligomer core minimizes the problem of end mismatch discrimination because the fixed core is internalized within the probe.

Degenerate probe ends provide an additional discrimination advantage. For a full match probe (i. e., all bases in the oligomer core match a sequence in the target DNA), most of the probe molecules in the pool will have one or a few mismatched bases among the degenerate base positions. For a single-mismatch probe, all probe molecules will have one additional mismatched base. Our research team and others [38] have shown that the discrimination observed between single-mismatch probes and those with two mismatches may be higher than that between full match and single-mismatch probes.

In addition to changes in probe design, hybridization conditions can maximize full match/mismatch discrimination and enhance detection of signals from short probes that form weak hybrids. Based on thermodynamic and kinetic considerations, Drmanac et al. (1990) predicted that the most favorable conditions for successful discrimination in hybridization experiments are low wash temperature, long wash times, and an excess of initial hybrid [39]. The researchers demonstrated experimentally that these conditions allowed accurate discrimination between full match and single-base mismatch probe/DNA pairs, thus confirming that general hybridization conditions may be used in most SBH experiments. In 1991, Strezoska et al. used selected octamer probes under such hybridization conditions to accurately assemble a 100-base DNA sequence. The redundant hybridization data obtained from the set of overlapping octamers easily compensated for low error rates observed in the experiment [32].

Differences in base composition can affect hybridization efficiency in probes of identical length [40], as would be expected based on the observed differences between delta G values of various dinucleotide pairs [41, 42]. Such effects can impact hybridization strength as much as changing probe length by several bases. A variety of experimental strategies have been tested to minimize this impact of base composition on hybridization efficiency [12]. These include using tetramethyl ammonium chloride for probes longer than 14 bases [43], changing probe length by adding one to four degenerate [19, 29, 31] or universal [44–46] bases, using appropriate base analogs [47], or simply adjusting hybridization conditions for different probe types.

Probe sequence, target/target hybridization [48, 49], stacking interactions [40], experimental imprecision and other factors influence absolute hybridization scores. Due to these factors the full match scores for some probes can be lower than the mismatch scores of others. Because of these uncertainties, control samples or groups of sequence-related probes [12, 29–32, 39, 50] are used to evaluate and transform absolute signals into relative values, which are then used in sequence determination.

In SBH experiments, actual hybridization signal data may be first used to distinguish positive (full match) from negative probes ("probe call"). These assignments are then used in the sequence assembly step to perform a "base call." However, an actual "probe call" step (assigning which probes are positive) is not necessary. A more general approach is to determine a score or probability that a given sequence is correct based on the scores of overlapped probes that fully match the suggested sequence vs. probes that have one or more mismatches for the same sequence.

2.6
Highly Parallel Data Collection in Different Formats

A hybridizing probe in an SBH experiment has completely random access to the data set (sequence), locating matching sequence strings anywhere within the target. This random access allows efficient parallel data processing, with the potential for thousands of probe-DNA hybridizations to be conducted simultaneously on a probe-or DNA-array or many other ways. One approach is to hybridize one labeled probe or probe sets at a time with arrays of samples [see, e. g., 12, 19]. Alternately, a single labeled sample can be hybridized with an array of probes [see, e. g., 12, 21, 22]. Furthermore, arrays can be designed in a predefined pattern of samples or probes or in a random collection of discrete particles that can be decoded at a later time [51]. Recent development of arrays on fiber optic bundles by Ilumina Inc. is very promising [55]. In addition to single component arrays, hybridization reactions can be performed with both components in solution in multi-well plates, especially if multiplex labeling of samples or probes is used [12]. Hybridization in solution requires efficient separation of hybrids from non-hybridized labeled components or a homogenous assay in which physically measurable changes occur upon hybridization, such as reduced fluorescence quenching when a probe hybridizes to DNA [53], or energy transfer between tags of two probes hybridized at a short distance [54]. The use of encoded micro or nano-tags and multi-well hybridization experiments will be discussed in a later section of this paper (see "Future Research Directions").

3
SBH Studies on Arrays of Samples

3.1
SBH on DNA Arrays

SBH methods are ideally suited to microarray technology due to their inherent potential for parallel sample processing. One promising SBH method involves exposing arrays of DNA samples to a single probe or probe pool [12, 19]. An important advantage of using a DNA array rather than a multiple probe array is that all resulting probe-DNA hybrids in any single probe hybridization are of identical sequence. This feature allows experimental conditions to be optimized for each hybrid type. In addition, using arrays of samples requires very simple sample preparation.

Arrays of samples are routinely used at Hyseq, Inc. and in other laboratories [55–58] to screen thousands of cDNA samples for gene identification. Selected sets of about 300 7-mer probes are used for such screening. Over 55,000 individual samples and controls can be spotted to a single 16×24 cm Genescreen membrane. Automated production lines are used to handle the large number of samples and hybridizations. For complete sequencing of 1–2 kb samples, approximately 10,000 probes or probe pools have to be tested, requiring 100 hybridization cycles on 100 replica arrays of DNA samples (see Sect. 3.3).

3.2
Blind De Novo Sequencing Test on Arrays of Samples

Drmanac et al. confirmed the accuracy of SBH in a blind test in which traditional gel methods and SBH were used to sequence three homologous primate T-cell DNA fragments [29]. The DNA targets, derived from human and rhesus monkey species, were 92–94% similar in sequence. The experiment used a pre-selected incomplete probe set that was carefully designed to test discrimination between full- and partial-match probes, and to distinguish very similar DNA targets from one another.

In this study, one research team sequenced three 2-kb T-cell loci by the dideoxy gel-sequencing method. This information was then used to assemble a list of 272 octamer probes that would theoretically allow a second team to accurately sequence 116 bp regions of these clones by SBH. The selected probes contained about twice as many non-matching as matching probes and also contained 8-mers matching a fourth primate species not included in the test.

The probes used in this experiment were actually pools of sixteen 10- and 11-mer probes containing two degenerate (variable) bases surrounding a specific 8-mer or 9-mer core. The longer probes increase hybrid stability, improving signal intensity and discrimination. Positive and negative DNA controls were also included, to allow calibration of the hybridization signals for the amount of DNA in each spot and to measure background signal.

The DNA sequences of the three unknown targets were precisely determined after hybridization with only 156 of the 272 probes. Both SBH and traditional gel methods gave identical sequence results for all target DNAs in the study. The average discrimination value (ratio of signals obtained from full match and single-base mismatch probes) was over ten in this study. False positives occurred infrequently and were generally due to probe hybridization with vector DNA or summed signals from two or more single-base mismatch probes.

3.3
Sample Arrays for Novel Polymorphism and Mutation Discovery

As genetic screening and gene therapy become increasingly important medical procedures, sequencing methods are needed that can rapidly and accurately detect mutations in patient DNA samples. Mutation detection by traditional gel sequencing methods is accurate but generally too slow and expensive for widespread use. In contrast, SBH using arrays of DNA samples provides an efficient, cost-effective way to screen large numbers of patients for individual genetic variations.

Drmanac et al. [31] confirmed the accuracy and speed of SBH arrays in detecting mutations in a medically important gene. The researchers used in situ mutagenesis to create a variety of insertion, deletion, and substitution mutations in a single p53 clone. PCR-amplified DNA samples from these mutants were then spotted directly to membranes without further purification, a protocol that could be applied to any patient DNA sample. Replica arrays were created containing all p53 samples and a variety of positive and negative controls.

In a blind experiment, 7-mer probe sets were then used to sequence all of the mutated p53 samples. The researchers used two probe sets in the p53 experiment, a complete set that matched all possible 7-mers and a smaller set that matched only the 7-mers present in the p53 reference sequence. Mutations were detected by comparing the average hybridization scores obtained for mutant and reference DNA samples using the reference set of probes. Figure 3 is a typical mutation-scanning graph, a plot of the ratio of hybridization strength of sample (S) vs control (C) DNAs as a function of base position. When test and control DNAs are identical in a given region, S/C ratios are expected to be one. When a mutation occurs in a test DNA, the S/C ratio for the mutation scanning graph drops dramatically due to poor hybridization of the reference probes with the mutant DNA. This situation is seen at base positions 143 and 157 in Fig. 3. The exact sequence in the mutation region is then determined by hybridizing the complete probe set.

A total of 13.2 kilobases of p53 DNA were sequenced with 100% accuracy in this experiment. The most difficult sequence assignments occur in regions where a single base is repeated several times in a row. When a point mutation occurs in one of these regions, only a few overlapped probes differ between the two sequence variants. Longer probes were used successfully in such cases.

One advantage of the SBH process when large sets of probes are used for mutation detection is that multiple, independent reads of both DNA strands help ensure accurate sequence determination. In this study, five replicate samples of one p53 gene were amplified and sequenced independently to confirm the reproducibility of the SBH process. The coefficient of variation for median S/C values

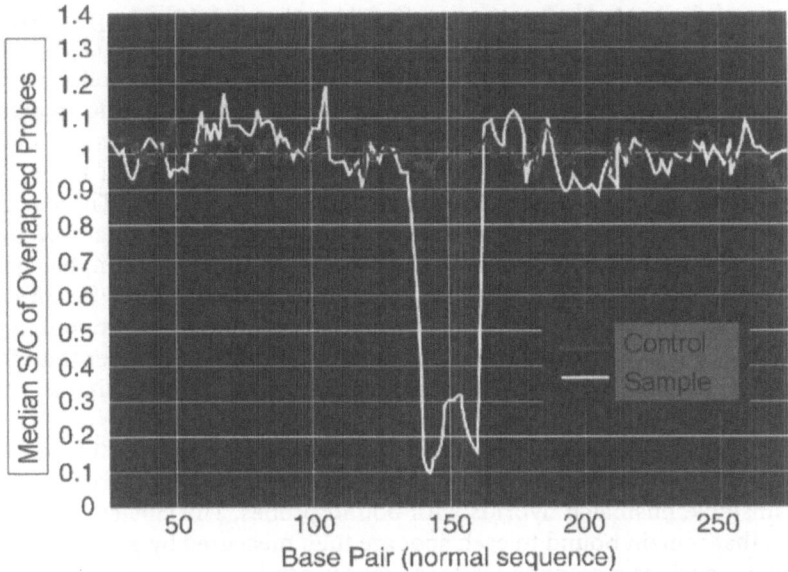

Fig. 3. A mutation-scanning graph showing two mutations in a p53 gene. Mutation scanning graphs are used to detect mutations in test samples relative to control (reference) DNAs. The S/C ratio represents the ratio of probe hybridization signals obtained for sample and control (reference) DNAs. For mutation scanning graphs the probe set used corresponds to the reference sequence. If the test sequence is identical to the reference, the median S/C ratio of seven overlapping probes will be close to 1, if a mutation is present in a given region, the median S/C ratio drops substantially because the reference matching probes do not bind to the mutated target DNA. In the mutation-scanning graph the flat line corresponds to a mutation-free control DNA; the sample line shows two mutations, a four-base deletion at position 143 and a G to A substitution at position 157

at mutant sites was below 3% for all cases, which is considered well within standards for most clinical diagnostic studies.

An identical experimental setup with 384 replica arrays on one membrane has been used successfully to genotype ApoE variants in samples from approximately 400 patients with various cardiovascular indications [59]. An accuracy of over 99% was achieved.

4
SBH on Arrays of Oligonucleotides

4.1
Initial SBH Tests on Probe Arrays

SBH analysis of a single complex nucleic acid sample may require tens of thousands of individual probe-DNA hybridizations. Such massively parallel experiments are ideally suited to microarray applications in which oligonucleotide probes bound to a solid support are exposed to DNA samples alone or in combination with nucleotide bases or additional probes in solution (see below). Probe arrays may be prepared by spotting pre-made probes [22, 60] on a properly prepared substrate [61] using a robotic station equipped with needles, capillaries or various non-contact pipetting devices such as piezoelectric controlled nozzles. Probes can also be positioned by electric force on a chip containing an array of individually addressable electrodes (Nanogen). Alternately, arrays can be prepared by in situ synthesis, in which the required bases are coupled sequentially using a variety of methods for spot-specific delivery of reagents. These include physical masking [21], ink-jet or bubble-jet printing nozzles (originally proposed by Southern [21], and now applied by Agilent, Protogen, Rosetta Inpharmatics, and Canon), photolithographic [62] or micromirror [63] methods for directing light on photosensitive nucleotide protection groups, and micro-stamping procedures [64]. In principle, quality control of spotted arrays is higher than for in situ synthesis, due to the very high quality synthesized probes used in their manufacture.

In one implementation of SBH, a labeled DNA sample is exposed to a probe array, which is then washed to remove target DNA that fails to hybridize or that forms unstable mismatch hybrids with bound probes. The labeled, hybridized samples that remain bound to each spot are then measured by a suitable detection device, such as a scanner. DNA sequences can be assembled from the hybridization results. The probe-DNA hybrids formed in such probe arrays have many different hybridizing sequences, which poses a technical challenge because pairs with different thermodynamic stabilities are formed and scored in a single hybridization reaction.

Several sequencing demonstrations involving probe arrays have been performed, typically using small arrays of selected overlapping probes to sequence short DNA targets. An *in situ* prepared array of 256 8-mer probes containing only A and C bases was used to sequence a 20-base target [65]. Mirzabekov et al. also used small spotted arrays of overlapping probes to sequence synthetic oligonucleotide targets [66], and recently prepared complete arrays of all 6-mers [67]. Gunderson et al. [68] developed an advanced format 2 SBH procedure, previously proposed by Charles Cantor [69], in which complete arrays of 8-mer and 9-mer probes are attached to double-stranded anchors. Target DNA fragments with ends complementary to the probe are then ligated to the array. Target DNA of 1 – 2 thousand bases were re-sequenced with low error rates using this highly redundant array, which provides over 100 measurements per DNA base [68]. Longer samples (5 – 16 kb) were also analyzed. In addition to data obtained on test samples, similar hybridization data obtained for reference samples of known sequence were used in the base-calling process [68].

4.2
Use of Non-Universal Arrays of Probes

Most current probe array SBH applications use non-universal sets of probes. Arrays prepared by in situ synthesis may contain 40 to 400 thousand probes. Such large arrays allow experimental designs featuring sets of probes suitable for large scale expression monitoring [70], SNP genotyping [71], or SNP/mutation discovery using a combination of full match, single-mismatch and single-base deletion probes matching each position in the reference sequence [72]. In SNP geno-typing studies, over 1000 different SNPs may be scored simultaneously. The very large number of long probes allows SNP and single-base mutation discovery in very long DNA targets, demonstrated by an analysis of the entire 16.6-kb mito-chondrial genome using an array of 135,000 15-mer probes [30]. In another ap-plication, an array of overlapped probes of varying lengths (up to 12 bases) was designed to match a series of tRNA sequences, to study how target secondary structures affect hybridization [49]. Further increases in the use and efficiency of non-universal probe arrays designed for specific sequences are expected.

4.3
Hybridization Combined with Polymerase

The use of DNA polymerase allows researchers to elongate already-hybridized probes, increasing the amount of sequence information obtained. Hybridized oligonucleotides may be extended by one or more bases using DNA polymerase and a set of four dNTPs tagged with different labels such as fluorescent tags [50, 69, 73, 74]. Deoxynucleotide bases must be sequentially blocked and unblocked in each elongation cycle, to prevent indiscriminate elongation of the probe. Each elongation cycle increases four-fold the number of oligonucleotide sequences that can be scored. Such sequence extension eliminates the need for initial probe labeling. Use of polymerase adds specificity at the extended 3' end, but the process may increase mismatches at the 5' end of probes.

4.4
Hybridization Combined with Ligase: Combinatorial Probe Scoring

Combinatorial ligation of two sets of short probes to create long probes is another powerful way to couple the hybridization reaction with an enzymatic or chemi-cal covalent bonding step [50]. This combinatorial sequencing approach has several advantages over traditional SBH, particularly the ability to detect long complementary sequences in target DNAs without the use of large sets of long probes. For example, two sets of 4096 6-mers can be combined to score over 16,000,000 12-mers. Only a small number of manufactured probes are needed ini-tially and each probe can be synthesized individually by standard chemical pro-cedures to create large quantities of very high quality probes for use in millions of assays. By requiring simultaneous initial hybridization of two short probes instead of a single long one, the overall hybridization specificity is increased substantially.

Covalent bonding provides a mechanism for accumulating sufficient signal to detect low-stability hybrids, which at any moment are present in very low quan-tities at the higher temperatures that are required to discriminate mismatches in very stable hybrids. Furthermore, such bonding helps minimize competition from target-target hybridization, a natural process that affects DNA interactions with probes. As covalent bonding of probes occurs, it removes target molecules from solution, driving target-target hybrids to dissolve to maintain equilibrium. Use of ligase also helps provide additional full match specificity, at least at the lig-ation site. In covalent bonding of adjacent probes, full match/mismatch discrimination is dependent in part on the equilibrium ratios of the corresponding hybrids during the hybridization reaction. Powerful low temperature discriminative washing procedures [39] cannot be used here to further increase specificity, but hot wash procedures after the ligation reaction can efficiently reduce background of non-specifically bound, labeled probes, thus increasing detection sensitivity.

One very effective implementation of this combinatorial sequencing approach involves the use of two sets of short probes (usually 4 to 10 bases long) – one set bound to a support and another labeled set in solution [50, 75, 76]. Both sets of probes are hybridized to target DNA samples in the presence of DNA ligase. Bound and labeled probes that hybridize to the target DNA at precisely adjacent positions are ligated to form a labeled, support-bound long probe. Probes that do not hybridize to DNA samples at adjacent sites do not ligate and are removed in the wash process. The labeled, support-bound long probes formed by ligation are then scored by an appropriate detection device.

The combinatorial SBH process has several unique advantages compared to simple probe array SBH experiments. Because a relatively small number of probes are used in each array, it is economical to create millions of identical arrays by spotting large quantities of high-quality probes prepared by standard synthesis. Furthermore, combinatorial SBH couples the capacity for spatial addressing inherent in probe arrays with the multiplex labeling advantages of probe pools in solution. This allows researchers to efficiently score millions of possible hybridization events for each sample using only a few thousand probes. For example, 10 – 100 subsets of differently labeled probes can be synthesized using a variety of fluorescent or mass tags. These subsets can then be combined to create pools that contain mixtures of 10 – 100 differently labeled probes, allowing independent scoring of 10 – 100 distinct hybridization events for each spot in the array. An additional advantage of using ligase-mediated covalent bonding of labeled probes to immobilized probes is the ability to use hot-wash procedures to conveniently and thoroughly eliminate background signal created by nonspecific adsorption of the labeled molecules to the support.

In summary, combinatorial SBH methods may use sets of hundreds to millions of short probes to test millions to billions of 10- to 20-mer (or longer) oligonucleotides in a DNA sample. The method provides highly efficient, robust, and cost-effective methods of sequencing complex DNA or RNA targets.

4.5
A Combinatorial SBH Blind Test

Drmanac et al. confirmed the specificity and sensitivity of combinatorial SBH in a blind sequencing test using complex synthetic DNA targets [77, unpublished results]. Replica arrays of fifty 5 – 7-mer probes were spotted on nylon membranes and then hybridized in the presence of T4 ligase with a variety of specially designed DNA targets and 20 – 30 radioactively labeled 5 – 7-mer probes. The study, performed in collaboration with the National Institute of Standards and Technology (NIST), resulted in correct determination of 15 out of 16 homozygote and heterozygote sequences in complex DNA targets comprising several kilobases.

Only a few overlapping probes were needed to sequence each site due to the high full match/mismatch discrimination observed in the experiment (an over 10-fold difference on average).

4.6
De Novo Sequencing, Comparative Sequencing and Genotyping Using Combinatorial Ligation and the HyChip System

Researchers at Hyseq, Inc. and Callida Genomics have developed the HyChip system, a universal combinatorial sequencing chip. The HyChip system consists of facing glass microscope slides containing four replica arrays of 1024 oligonucleotide probes consisting of all possible 5 base combinations. These bound 5-mer arrays are then exposed to DNA samples and a complete set of 1024 TAMRA-labeled 5-mers, combined in various pre-mixed probe pools containing 16 to 256 probes per pool (Fig. 4).

With the combinatorial advantages of its ligation process, the HyChip universal system can score a complete set of over one million 10-mer probes per sample using only 2000 5-mer probes. DNA samples are generally prepared by PCR from genomic DNA or various clones. The complete probe sets used allow universal genetic analysis of any gene or DNA sample from any organism. In addition, the HyChip system allows efficient analysis of DNA samples of different length. Hyseq researchers have shown that a large number of probes with a single label type can be hybridized together as one pool without loss of information as long as relatively short DNA targets are used. When using complete sets of probes, about 4–8 hybridization scores are preferred per base of target DNA analyzed. To analyze 1000 bp of target DNA usually requires at least 4000 hybridization scores. This result can be obtained using 4 pools, each containing 256 labeled 5-mers. After adding target DNA to each pool the 4-pool set is tested on one HyChip having four replicas of 1024 immobilized probes, producing 4096 measurements. Probe pools work effectively, especially for short DNA targets, because only a fraction of fixed probes within the array give a positive result for each pool. In the case of a 1000 bp single stranded nucleic acid target sequenced using 256-probe pools, less than 1000 10-mers are full match, thus the remaining 3000, of 4000 pool scores, are negative.

All major types of DNA analyses have been demonstrated using the HyChip system. De novo sequencing, the most challenging computational process, has been performed successfully using DNA targets of 100–1000 bases. The best results were observed with targets of about 500 bases in length. In a simulation experiment using 16,384 pools of 1024 12-mers, we demonstrated a 90% or greater success rate in assembling 3200 bp sequences. This result confirms that efficient de novo sequencing can be achieved with as few as five hybridization scores per base (16,384/3200) [25] using DNA targets that are over three times as long as those typically sequenced by traditional gel methods.

Hyseq researchers have also performed comparative sequencing, re-sequencing and SNP discovery experiments on several human, bacterial, and viral (e.g., HIV) genes. Reference sequences, although not required, were used to aid in the assembly steps of these applications. It was, however, not necessary to obtain

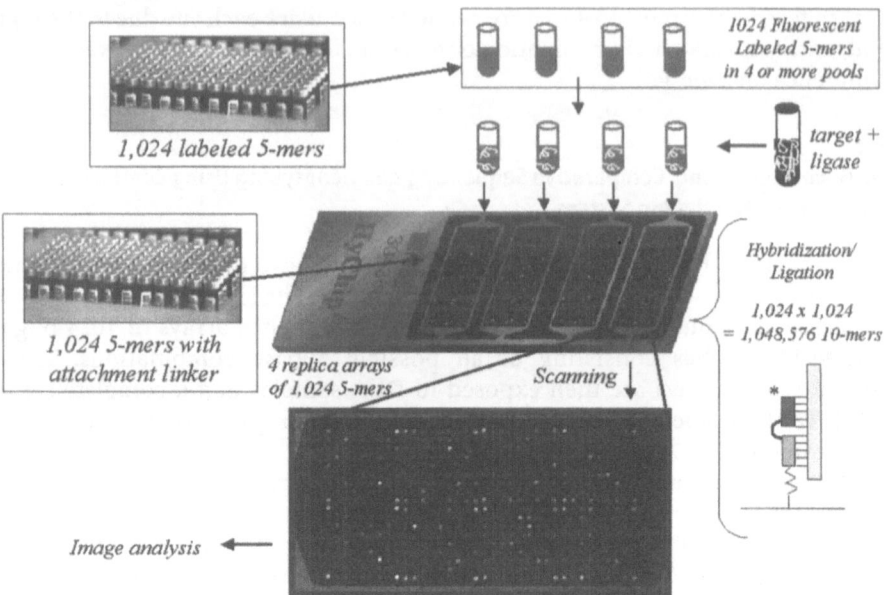

Fig. 4. Universal HyChip system: components and process. The system is composed of pentamer probe arrays on glass microscope slides and pools of TAMRA-labeled pentamers. Each slide has four replica arrays of 512 pentamers separated by a hydrophobic mask. Matching pairs of slides with different 512 probe arrays are sandwiched to form four capillary hybridization chambers, each chamber containing a complete set of 1024 pentamer probes. Arrays are prepared by spotting pre-synthesized probes (each coupled with a spacer and a primary amine group at its 5′ end) using an 8-channel spotting device. The probes are spotted as 200 um dots on glass slides that have been derivatized with 3-propylamino silane and with 1,4-phenylene diisothiocyanate. A complete set of 1024 TAMRA-labeled probes stored in vials is pipetted to create sets of 4, 8, 16, 32, or 64 pools containing 256 to 16 probes respectively. Each pentamer probe is present in only one pool in a given pool set. Target DNA, T4 ligase and buffer are added to each pool and the mixture is loaded by pipette onto the HyChip system, where it is drawn by capillary forces into a selected hybridization chamber. The hybridization/ligation reaction is performed for about one hour, allowing accumulation of positive 10-mer probe constructs. Each 10-mer formed consists of one labeled probe that has been ligated to an immobilized probe using the DNA target as a template. The presence/absence of each of the 1,048,576 possible 10-mers in the sample is thus tested in this combinatorial scheme. After incubation, the slides are hot-washed to remove unattached, labeled probes and DNA, and then read on commercial laser scanners (Axon, Luminex, Vertek). Hybridization results are then scored by image analysis software. A pattern of positive and negative control spots is visible in the figure

and use hybridization data from reference samples themselves, thus simplifying and reducing the cost of the sequencing process. Amplicons ranging in size from 100 – 10,000 bases in length were tested using sets of probe pools containing 16 to 256 labeled probes. By combining hybridization/ligation data for two complementary strands researchers achieved nearly 100% sequence accuracy in SNP discovery and complete re-sequencing studies of targets containing the sort of compound multiple mutations typically found in HIV samples.

Sequencing complementary DNA strands independently minimizes the impact of pool-related false positive probes because the real positive probes for each complementary strand tend to fall, by chance, in different pools. Thus, the majority of real positive probes test positive in both experiments, while the majority of pool-related false positive probes test positive in only one experiment. We also found that sequence regions that produce low positive signals in one strand usually produce much stronger signals in the complementary strand, presumably because G/T-stable mismatches that cause target-target interactions in one strand correspond to very unstable C/A-mismatches in the complementary strand. In addition, having two data sets reduces experimental noise by providing twenty scores per base pair (10 per strand).

An important advantage of redundant probe measurements per base is the ability to improve the accuracy of each base assignment or to declare some bases uncalled. With current non-optimized probe sets of uniform length and design (which produce hybrids of very different stability) single-strand sequencing generally result in about 2–3% uncalled bases. These are typically located in a few DNA regions, and they are almost always resolved when data from complementary strands or overlapping targets are included.

The use of reference sequences to guide analysis of hybridization data allows researchers to process much longer DNA samples than can be handled in a *de novo* sequencing experiment. Using combinatorial SBH we have sequenced the entire single-stranded 5.4 kb genome of bacteriophage PhiX174, confirming the power of the SBH method to read much longer DNA molecules than can be processed by any other sequencing method today. In this experiment, 32 probe pools containing 1024 probes were used to generate about 33,000 individual measurements, or roughly 6 measurements per base. Furthermore, the HyChip system's capacity to analyze complex DNA allows researchers to sequence mixtures of separate, unrelated DNA fragments, a unique advantage of the SBH process. This feature was confirmed by successfully sequencing a complex mixture of HIV, p53 and ApoB amplicons using the HyChip system.

Hyseq researchers have also demonstrated efficient sequencing of several DNA targets that are AT rich (<40% G+C) or GC rich (>70% G+C). Some of these targets have proven very difficult to sequence by the Sanger dideoxy method due to stops or compressions in the sequencing process.

Accurate identification of heterozygotes is also routinely achieved with data obtained from the HyChip system, as shown in Fig. 5. Interestingly, both correct bases at the heterozygous site scored as high as did single bases in homozygous sites. The high signals observed for both alleles allow separation from background noise and correct identification of the heterozygous bases.

Another application of the universal HyChip system is genotyping, which involves detection of known allele sequences in a DNA sample, and is usually done by use of locus-specific primers and probes. In SBH genotyping studies, short DNA samples containing a polymorphic site are prepared individually or simultaneously in a multiplex PCR reaction. A single HyChip system with four labeled probe pools is sufficient to genotype 10–20 sample amplicons roughly 50 base pairs in length, providing the preferred 4 to 8 scores per base for the combined 500–1000 bases. Using the HyChip system, each amplicon is completely rese-

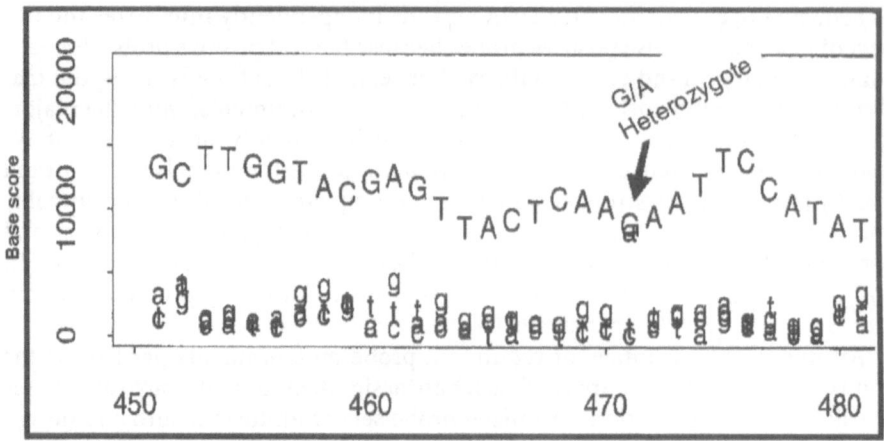

Base position

Fig. 5. Sequencing a heterozygote site in the ApoB gene using the HyChip system. The base-calling graph of a region from 450–480 base position of a 700 bp ApoB amplicon is depicted. The base scores represent the average relative fluorescent signal of ten overlapping probes corresponding to each of four possible bases at each position. Capital letters represent reference bases of the ApoB gene. At every base position except 471, the three non-reference bases have a much lower score than that of the reference base. At base 471 both A and the expected reference G show almost identical high scores, with the other two bases showing low scores typical of an incorrect base

quenced, with each base in the polymorphic site (homozygote or heterozygote) read by 10 overlapping probes. Ligation of two complete sets of short probes offers advantages of universality, redundant testing of each allele with multiple overlapping probes, and eliminates the need for new probe design and test optimization with each new SNP.

An additional option for genotyping is to use universal arrays and to select from a universal set of labeled probes a few probes corresponding to each polymorphic site. In this case a single pool of 128 labeled probes can test 32 SNPs, with four overlapping labeled probes corresponding to each locus. One HyChip system can be used to test a set of 32 SNPs in four individuals, or, using four specific probe pools, to test a set of 128 SNPs in a single sample.

In summary, combinatorial SBH based on the universal 5-mer HyChip system has great potential for sequencing, genotyping, and polymorphism and mutation discovery. A universal 5-mer HyChip system could be used to monitor expression of a selected set of dozens of genes, or, if enlarged to 6–8-mer capacity, to sequence over 100 kb in one reaction, and to monitor gene expression. The HyChip combinatorial system allows testing of complete sets of millions of possible oligonucleotides in each sample, resulting in long, accurate DNA reads, and providing the universality and flexibility to analyze any sample, any gene, or any sequence variant.

5
Future Research Directions

5.1
Hybridization Technologies

In SBH experiments, analysis of increasingly complex DNA molecules generally requires use of longer probes and larger probe sets. Practical analysis of long targets with complete probe sets requires that the probes chosen be long enough and pools small enough that less than half of the probes or pools tested result in positive hybridization. A DNA molecule of length L bases has approximately L positive probes. This imposes a minimum of 2L measurements (two per base); at least 50% of which should be negative. Note that the number of overlapping probes that read each base is equal to the number of bases in the probe, independent of target length and number of probes per pool. In general, hybridization data from four to ten probes (or probe pools) per base will yield proper assembly and high accuracy in base calling. By comparison, single-pass gel sequencing also requires four or more "scores" per base.

A variety of different techniques can be used to further increase the throughput of SBH. Use of multiple probe or DNA labels may greatly simplify SBH experiments. For example, sequencing 1 Mb of DNA in a single reaction may use up to 10 million hybridization measurements. These may be obtained with a 10 million-element array, or with a 1 million-element array if ten different fluorescent labels are used in a combinatorial scoring approach (see Sect. 4.4). The required number of DNA or probe spots may be further reduced if a highly multiplex system of 100 or more tags is used, for example, by use of mass tags that can be identified by mass spectroscopy [79]. Currently, probe arrays or random arrays of micron size beads on optical fibers (Ilumina Inc.) may have effective feature spans of approximately 15 microns, permitting roughly 500,000 features per square centimeter. At such a density, sequencing an entire bacterial genome at once using a single label would require a large 10×10 cm chip. Further reduction of feature size below 5 um would greatly enhance the potential of SBH arrays.

Non-array methods may also be used to efficiently score large numbers of probes. Color-coded or otherwise distinguishable micro-beads may be used [51]. For example, Luminex has developed a flow-sorting detection system with 100 distinct beads encoded by specific ratios of two fluorescent dyes. In this system, a single small well containing 100 beads would replace 100 DNA or probe spots, a 1536 well plate would replace an entire array of 153,600 single color spots, and 60 plates would permit the 10 million hybridization measurements needed for analysis of 1 Mb of DNA. The development of multiplex systems containing 1000 or more encoded beads would obviously be of great benefit [51]. It may be possible to produce several thousand unique Q-beads (QuantumDot, Inc.), and virtually unlimited numbers of micro transponders with specific emission frequencies can be designed (PharmaSeq Inc.). Unfortunately, current micro-transponders are still quite large (250 um), and may ultimately prove to be too expensive for most research purposes. Another new SBH platform combines micro-fluidics with probe arrays (FeBiT, Inc., Motorola, Inc.) or bead assemblies [80] located in micro-capillary channels.

The oligonucleotide probe scoring process may ultimately be miniaturized to involve detection of a single positive probe molecule in solution. Such a process would provide virtually unlimited multiplexing capabilities [81]. A variant of this probe labeling and detection process is schematically presented in Fig. 6.

Nano-barcode tags are prepared as chains of nano beads, other particles, or large molecules. An example with four different monomer units (combination of two lengths and two diameters) is depicted in Fig. 6. In this case one of four monomers is assigned to each base. To encode all N-mer probes, the barcode chain has to be N elements in length. Sets of one million 10-mers to one trillion 20-mers can be encoded and potentially analyzed in a single reaction. After hybridization, probes bound to the target may be separated from unbound probes and their barcodes read by the detector device. The advantage of three-dimensional nano-tags is the ability to identify single molecules when a tagged molecule passes between a pair of nano-electrodes. The flow of electrons between electrodes will be influenced differentially by the size and shape of different monomer tags, providing a specific signature.

Large sets of tagged probes can be prepared in a sequential mix-and-divide synthesis scheme. An equimolar mix of four tagged monomers is split into four separate synthesis reactions to add A, T, C, and G to the oligonucleotide end and a corresponding base-specific tag on the tag end, resulting in 16 dinucleotides, each labeled with a specific dimer tag. By repeating n cycles of mixing and di-

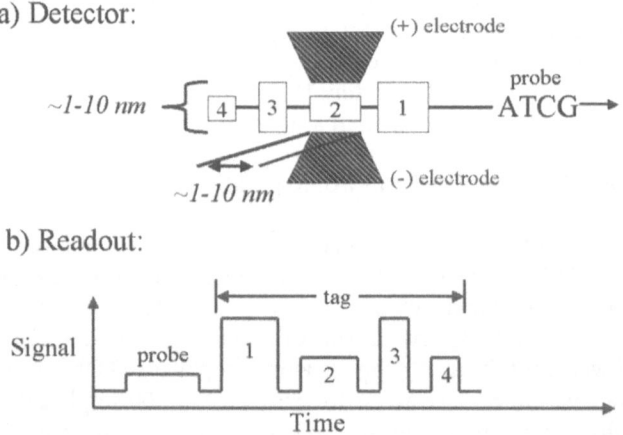

a) Detector:

b) Readout:

Fig. 6. SBH in solution using three-dimensional nano-barcode tags. **a** Probe ATCG tagged with a four-monomer tag passes between a pair of closely spaced 1–10 nm electrodes. **b** Readout of signal variation over time. The signal graph reflects length and diameter of probe molecule as a whole, without ability to distinguish individual bases, which are only about 0.3 nm in diameter. Alternately, the individual barcode tag units cause significant differences in strength or duration of the electronic signal. By using sets of four monomer tags, all possible n-mer probes can be identified by n-unit-long barcodes. Tens of millions of tagged probes can be prepared with only a few dozen coupling reactions in a mix-and-divide synthesis scheme

viding while adding new base/tag pairs, all possible n-mer probes can be prepared in a series of 4n syntheses. Construction of all possible 20-mers would thus require 80 reactions. As with other SBH procedures, nano-tagged probes can be hybridized directly or extended by use of DNA polymerase. In addition, two mixtures of shorter probes (tagged at opposite 5' and 3' ends) can be ligated and efficiently analyzed by reading the two tags at the ends of the ligated oligonucleotides.

Other researchers have also proposed the possibility of distinguishing natural bases directly using a similar principle, by passing DNA through membranes with nano-pores [8]. The main difficulties facing this process are the very small sizes of the nucleotide bases and the slight variations in size that exist between A and G purines and T and C pyrimidines. Nanotags allow engineering of monomers from materials that are several times larger than natural bases, with large differences in monomer size to assure quick and accurate detection.

Development of SBH technology with single molecule sensitivity can lead to whole human genome analysis on a single tiny device, without need for DNA amplification or fractionation. The ability to rapidly read billions of tags or bases in parallel analyses using thousands of reading channels would facilitate large genome sequencing in a single day.

5.2
Hybridization Chemistry

The hybridization success of full match probes depends primarily on the differential stability of probe/target and target/target hybrids. Probes which form less stable hybrid duplexes not only produce inherently weaker signals, but are also less likely to compete with imperfect but longer target/target hybrids. Probe designs that increase hybrid stability can dramatically improve hybridization efficiency. For example, lengthening a probe by one base may produce a 10–20-fold stronger signal than can be achieved with a 10-fold higher concentration of the shorter probe (unpublished data). The challenge is to obtain optimal stability for all probes under the same conditions, because low efficiency probes do not produce strong enough full match signals, while probes with very stable full match hybrids also produce strong signals with mismatch targets.

The ideal design solution would be a set of probes with the following characteristics:

1. equal hybridization stability among all possible full match probe-DNA pairs,
2. probe-DNA hybridizations that are stronger than competing DNA-DNA or probe-probe hybridization (even for DNA-DNA hybrids longer than probes), and
3. high full match/mismatch discrimination for all types of mismatches in all sequence contexts.

Both peptide nucleic acids (PNA) [82, 83] and locked nucleic acids (LNA) [84] may increase hybrid stability, and have great potential to meet these conditions, unless non-standard binding produces an intolerably high level of artifacts. Unfortunately, the modified chemical backbone of such probes may limit use of enzymes (such as polymerase, ligase or mismatch recognition enzymes) used for enhanced hybridization scoring.

There are several ways to equalize the hybrid stability of DNA (or PNA and LNA) probes of differing base composition. These methods include the use of longer informative sequences to increase the binding strength of (A+T) rich probes, or the use of variable numbers of degenerate bases at the probe ends [19, 31]. Development of universal bases that form pairs with delta G values equal to those of regular A/T pairs would be desirable. Current universal bases [44, 45] usually pair with natural bases with binding efficiencies comparable to mismatch base pairs. One intriguing possibility would be to combine an LNA-modified sugar with an existing universal base to enhance the binding efficiency. In addition, nucleotide base analogs may be used, particularly to increase the stability of A/T pairs. Promising new options are:

1. The use of LNA-modified T or A bases, coupled with use of standard C/G bases, (which naturally bind more strongly than A and T), and
2. coupling special minor groove binders to A+T rich probes to stabilize their hybridization to complementary DNA [85].

On average, probes with roughly equal hybrid stability give similar discrimination values. To further minimize mismatch signals, especially of G/T pairs in certain surrounding sequences [36, 37], potential use of electronic stringency, [86] or mismatch recognition and/or cutting enzymes [87, 88] or chemicals [89] may be beneficial. In the case of combinatorial probe scoring (e.g. HyChip), use of selected [90] or specially designed ligase enzymes that are highly sensitive to mismatches within several bases of the ligation site may provide a solution that does not require any additional treatment.

5.3
Advanced SBH Applications

There are several important research and medical applications that demand extremely efficient and accurate DNA sequencing. These include:

- Routine research sequencing of thousands of new bacterial and eukaryotic genomes,
- Cataloging of rare genetic polymorphisms, by sequencing thousands of individual human genomes,
- Complete resequencing of viral and bacterial genomes to identify drug resistance or virulence genes for infectious disease management,
- Genetic screening and diagnostics, via complete sequencing of hundreds or thousands of the most medically important genes in infants, cancer or heart disease patients, etc., and
- Complete genome analysis for medical prognostics and diagnostics, by genotyping millions of SNPs or completely re-sequencing the genomes of all newborn infants.

Achieving these research and medical needs may require development of sequencing methods that can acquire over 1 GB of highly accurate, inexpensive sequence data per instrument per day. This throughput capacity, which is 1000 – 10,000-fold higher than currently available, will most likely require the use of integrated micro-fluidics to process genomic DNA samples without handling millions of amplified fragments in multi-well plates. The inherent parallelism and long read-length of SBH technology may be the only practical solution to these challenges.

6
Conclusion

We live in a world of self-encoded, evolving biological systems, each of which possesses precise instructions for development and function recorded in its genomic DNA or RNA. Advances in the medical, biological and agricultural sciences are thus critically dependent on obtaining a large amount of accurate DNA and RNA sequence information quickly at the lowest cost. Despite great progress, current sequencing methods still limit the acquisition of and realization of the value of these large sequence data sets in modern research. Even though the entire human genome has been successfully sequenced there remains a great need for advanced new sequencing technologies. Efficient methods such as SBH have the potential to obtain accurate DNA sequence information at 1000-fold faster rates than are achieved today. This ultra-fast DNA sequencing would propel progress in the life sciences, with tremendous potential to improve overall human health and well-being.

Acknowledgement. We thank Kathy Ashue, Rodney Bellot, Haibin Chen, Linsu Chen, Linh Duong, Jenny Le, Charlene Lee, Halina Loi, Xinyu Mu, Kiran Mukhyala, Thutrang Nguyen, Helena Perazich, Khao Pham, Tami Pham, Tran Van, Christine Wong, Wei Wu, Linda Wu, Sean Sleight, Pouneh Tavakoli Han Wang and Sherry Zhang for their assistance throughout this project and Elizabeth Garnett for her organizational efforts and preparation of figures.

7
References

1. Sanger F, Nicklen S, Coulson A (1977) Proc Natl Acad Sci USA 74:5463
2. Hyman RD (1988) Anal Biochem 158:423
3. Ronaghi M (2001) Genome Res 11:3
4. Jett JH et al. (1989) J Biomol Struct Dyn 7:301
5. Lindsay SM, Philipp M (1991) Genet Anal Tech Appl 8:8
6. Beebe TP, Wilson TE, Ogletree DF, Katz JE, Balhorn R, Salmeron MB, Siekhaus WJ (1989) Science 243:370
7. Woolley AT et al. (2000) Nat Biotechnol 18:760
8. Church et al. (1998) US Patent 5 795 782
9. Sanger F, Brownlee GG, Barrell, BG (1965) J Mol Biol 13:373
10. Holey RW, Apgar J, Everett GA, Madison JT, Marquisee M, Merrill SH, Penswick JR, Zamir A (1965) Science 147:1462
11. Murray K (1970) Biochem J 118:831
12. Drmanac R, Crkvenjakov R (1987) Yugoslav Patent Application 570/87 (Issued as R Drmanac, R Crkvenjakov, Method of sequencing of genomes by hybridization with oligonucleotide probes. U.S. Patent 5,202,231 (1993)
13. Drmanac R, Drmanac S, Little D (2000) Encyclopedia of Analytical Chemistry: "Sequencing and Fingerprintig DNA by Hybridization", p. 5232 – 5257

14. Doty P, Marmur J, Eigen J, Schildkraut CE (1960) Proc Natl Acad Sci USA 46:461
15. Beaucage JL, Caruthers MH (1981) Tetrahedron Lett 22:1859
16. Wallace RB, Shaffer J, Murphy RE, Bonner J, Hirose T, Itakura K (1979) Nucleic Acids Res 6:3543
17. Poustka A, Lehrach H (1986) Trends Genet 2:174
18. Mullis KB, Faloona FA (1987) Methods Enzymol 155:335
19. Drmanac R, Labat I, Brukner I, Crkvenjakov R (1989) Genomics 4:114
20. Bains W, Smith GC (1988) J Theor Biol 135:303
21. Southern E (1988) International Patent Application PCT GB 89/00460
22. Lysov YP, Florentiev VL, Khorlyn AA, Khrapko KR, Shick VV, Mirzabekov AD (1988) Dokl Akad Nauk SSSR 303:1508
23. Macevicz SC (1989) International Patent Application PCUS89 04741
24. Pevzner PA, Lipshutz RJ (1995) Towards DNA sequencing chips. In: Privara I, Rovan B, Ruzicka P (eds) Mathematical foundations of computer science 1994 in: The proceedings of 19th international symposium, MFCS '94. Springer-Verlag, Kosice, Slovakia, Berlin
25. Drmanac R (1999) US Patent application published as WO 00/40758
26. Preparata FP, Fieze AM, Upfal E (1999) 3rd Annual International Conference on Computational Biology. Lyon, France
27. Khrapko KR, Lysov YP, Khorlyn AA, Shick VV, Florentiev VL, Mirzabekov AD (1989) FEBS Lett, 256:118
28. Drmanac R, Crkvenjakov R (1992) International Journal of Genome Research 1:59
29. Drmanac R, Drmanac S, Strezoska Z, Paunesku T, Labat I, Zeremski M, Snoddy J, Funkhouser WK, Koop B, Hood L, Crkvenjakov R (1993) Science 260:1649
30. Chee M, Yang R, Hubbell E, Berno A, Huang XC, Stern D, Winkler J, Lockhart DJ, Morris MS, Fodor SPA (1996) Science 274:610
31. Drmanac S, Kita D, Labat I, Hauser B, Burczak J, Drmanac R (1998) Nat Biotechnol 16:54
32. Strezoska Z, Paunesku T, Radosavljevic D, Labat I, Drmanac R, Crkvenjakov R (1991) Proc Natl Acad Sci USA 88:10089
33. Uhlenbeck OC, Martin FN, Doty P (1971) J Mol Biol 57:217
34. Doel MT, Smith M (1973) FEBS Lett 34:99
35. Dodgson JB, Wells RD (1977) Biochemistry 16:2367
36. Allawi HT, SantaLucia J Jr (1998) Biochemistry 37:2170
37. Allawi HT, SantaLucia J Jr (1998) Nucleic Acids Res 26:2694
38. Guo Z, Liu Q, Smith LM (1997) Nat Biotechnol 15:331
39. Drmanac R, Strezoska Z, Labat I, Drmanac S, Crkvenjakov R (1990) DNA Cell Biol 9:527
40. Wetmur JG (1991) Crit Rev Biochem Mol Biol 26:227
41. Breslauer KJ, Frank R, Blocker H, Marky LA (1986) Proc Natl Acad Sci USA 83:3746
42. Sugimoto N, Nakano S, Yoreyama M, Hanada K (1996) Nucleic Acids Res 24:4501
43. Wood WI et al. (1985) Proc Natl Acad Sci USA 82:1585
44. Berger M, Wu Y, Ogawa AK, McMinn DL, Schultz PG, Romesberg FE (2000) Nucleic Acids Res 28:2911
45. Seela F, Debelak H (2000) Nucleic Acids Res 28:3224
46. Fotin A, Drobyshev A, Proudnikov D, Perov A, Mirzabekov A (1998) Nucleic Acids Res 26:1515
47. Weiler J, Gausepohl H, Hauser N, Jensen ON, Hoheisel JD (1997) Nucleic Acids Res 25:2792
48. Doty P, Boedtker H, Fresco JR, Haselkorn R, Litt M (1959) Proc Natl Acad Sci USA 482
49. Mir KU, Southern EM (1999) Nat Biotechnol 17:788
50. Drmanac R (1993) U.S. Patent Application published as PCT/US94/10945
51. Drmanac R, Crkvenjakov R (1990) Scientia Yugoslavica, 16:99
52. Michael KL, Taylor LC, Schultz SL, Walt DR (1998) 70:1242
53. Gisendorf BAJ, Vet JAM, Tyagi S, Mensink EJMG, Trijbels FJM, Blom HJ (1998) Clin Chem 44:482
54. Chen X, Livak K, Kwok PY (1998) Genome Res 8:549
55. Drmanac S, Stravropoulos NA, Labat I, Vonau J, Hauser B, Soares MB, Dramanac R (1996) Genomics 37:29

56. Milosavljevic A, Zeremski M, Strezoska Z, Grujic D, Dyanov H, Batus S, Salbego D, Paunesku T, Soares MB, Crkvenjakov R (1996) Genome Res 6:132
57. Meier ES, Lange J, Gerst H, Herwig R, Schmitt A, Freund J, Elge T, Moss R, Herrmann B, Lehrach H (1998) Nucleic Acids Res 26:2216
58. Drmanac R, Drmanac S (1999) Methods Enzymol 303:165
59. Drmanac S, Heilbron DC, Pullinger CR, Jafari M, Gretzen D, Ukrainczyk T, Cho MH, Frost PH, Siradze K, Drmanac RT, Kane JR and Malloy MJ (2001) Journal of Cardiovascular Pharmacology and Therapeutics 6(1):47
60. Ekins R, Chu FW (1999) Trends Biotechnol 17:217
61. Beier M, Hoheisel JD (1999) 27:1970
62. Fodor SPA, Read JL, Pirrung MC, Stryer L, Lu AT, Solas D (1991) Science 251:767
63. Singh-Gasson S, Green RD, Yue Y, Nelson C, Blattner F, Sussman MR, Cerrina F (1999) Nat Biotechnol 17:974
64. Lu ZH, Zhao YJ, He NY, Sun X (2000) Intl Forum on Biochip Technologies Beijing, China
65. Southern EM, Maskos U, Elder JK (1992) Genomics, 13, 1008
66. Yershov G, Barsky V, Belgovskiy A, Krillov E, Kreindlin E, Ivanov I, Parinov S, Guschin D, Drobyshev A, Dubiley S, Mirzabekov A (1996) Proc Natl Acad Sci 93:4913
67. Proudnikov D, Kirillov E, Chumakov K, Donlon J, Rezapkin G, Mirzabekov A (2000) Biologicals 28:57
68. Gunderson KL, Huang XC, Morris MS, Lipshutz RJ, Lockhart DJ, Chee MS (1998) Genome Res 8:1142
69. Broude NE, Sano T, Smith CL, Cantor CR (1994) Proc Natl Acad Sci USA 91:3072
70. Lockhart DJ, Dong H, Byrne MT, Follettie MT, Gallo MV, Chee MS, Mittmann M, Wang C, Kobayashi M, Horton H, Brown EL (1996) Nat Biotechnol 14:1675
71. Wang DG, Fan JB, Siao C, Berno A, Yang P, Sapolsky R, Ghanour, G, Perkins N, Winchester E, Spencer J, Kruglyak L, Stein L, Hsie L, Topaloglu T, Hubell E, Robinson E, Mittman M, Morris MS, Shen N, Kilburn D, Rioux J, Nusbaum C, Rozen S, Hudson T, Lipshutz R, Chee M, Lander E, (1998) Science 280:1077
72. Hacia JG, Brody LC, Chee MS, Fodor SPA, Collins FS (1996) Nat Genet 14:441
73. Nikiforov TT, Rendle RB, Goelet P, Rogers YH, Kotewicz ML, Anderson S, Trainor GL, Knapp MR (1994) Nucleic Acids Res 22:4167
74. Canard B, Sarfati RS (1994) Gene 148:1
75. Drmanac R, Drmanac S (2001) Sequencing by hybridization arrays. In: Rampal J (ed) Methods in molecular biology. DNA Arrays: Methods and Protocols Humana Press, Totowa, NJ
76. Drmanac R, Drmanac S (2001) DNA sequencing by hybridization with arrays of samples or probes. In: Rampal J (ed) Methods in molecular biology. DNA Arrays: Methods and Protocols Humana Press, Totowa, NJ, 170:173
77. Drmanac S et al. (in preparation)
78. Guo Z, Guilfoyle RA, Thiel AJ, Wang R, Smith LM (1994) 22:5456
79. Graber JH, Smith CL, Cantor CR (1999) Genet Anal 14:215
80. Drmanac R (1998) US Patent application published as WO 99/60170
81. Drmanac R (1999) US Patent application published as WO 00/56937
82. Nielsen PE, Egholm M, Berg RH and Buchardt O (1991) Science 254: 1497
83. Nielsen PE, Egholm M, Buchardt O (1994) Bioconjug Chem 5:3
84. Wahlestedt C et al (2000) Proc Natl Acad Sci 97:5633
85. Kutyavin IV et al. (2000) Nucleic Acids Res 28:655
86. Radtkey R et al. (2000) Nucleic Acids Res 28
87. Lu AL, Hsu IC (1992) Genomics 14:249
88. Lishanski et al. (1994) Proc Natl Acad Sci USA 91:2674
89. Dianzani I, Camaschella C, Saglio G, Forrest SM, Ramus S, Cotton RG (1991) Genomics 11:48
90. Housby JN, Southern EM (1998) Nucleic Acids Res 26:4259

Received: June 2001

Protein Array Technology:
The Tool to Bridge Genomics and Proteomics

Holger Eickhoff[1] · Zoltán Konthur[2] · Angelika Lueking[2] · Hans Lehrach[2] ·
Gerald Walter[2] · Eckhard Nordhoff[1] · Lajos Nyarsik[2] · Konrad Büssow[2]

[1] Scienion AG, Volmerstr. 7b, 12489 Berlin, Germany. *E-mail: eickhoff@scienion.com*
[2] Max Planck Institute of Molecular Genetics, Ihnestrasse 73, 14195 Berlin, Germany

The generation of protein chips requires much more efforts than DNA microchips. While DNA
is DNA and a variety of different DNA molecules behave stable in a hybridisation experiment,
proteins are much more difficult to produce and to handle. Outside of a narrow range of environmental conditions, proteins will denature, lose their three-dimensional structure and a lot
of their specificity and activity. The chapter describes the pitfalls and challenges in Protein Microarray technology to produce native and functional proteins and store them in a native and
special environment for every single spot on an array, making applications like antibody profiling and serum screening possible not only on denatured arrays but also on native protein
arrays.

Keywords: Protein Array, Expression Profiling, Automation, Serum Screening, Antibody Profiling, Protein Purification

1	Introduction	104
2	From 2D Electrophoresis and Microtitre Plates to Microarrays of Biomolecules	105
3	Requirements for Protein Arrays	106
4	Planar Immobilisation of Proteins	107
5	Detection of Molecular Interactions on Microarrays	108
6	Applications of Protein Arrays	108
7	Outlook	109
8	References	110

Advances in Biochemical Engineering/
Biotechnology, Vol. 77
Managing Editor: T. Scheper
© Springer-Verlag Berlin Heidelberg 2002

1
Introduction

Each cell of a living organism contains the whole genetic information in the form of DNA molecules. The available golden path sequence of the human genome is known to be 3×10^9 nucleotide base pairs in size, coding for a currently unknown number of genes. Although the DNA information is, with rare exceptions, identical in each cell, several hundreds of different known cell types do exist. The simplified view of a cell brings up specific populations of DNA and mRNA molecules that are translated into cell-specific populations of proteins. Although this is only a simplified representation, we are currently able to understand only a minority of the complex interactions in the cell machinery comprising DNA, RNA and proteins as major compounds. Genomic databases have enabled us to access, retrieve and process biological information. The determination of the genomic sequences of higher organisms, including humans, is now attainable, but represents only one level of genetic complexity. The determination of the expression profiles and protein profiles of certain cell types represents the next level of genomic complexity, equally important to sequencing.

The interest of our research is focussed in finding all genes, their in vivo functions and the features of the corresponding proteins. Information about a gene's expression is important for its potential exploitation. A gene's expression and the corresponding protein level can be highly specific to a tissue, organ, cell type or disease and, as such, may be attractive as targets for the development of highly specific therapeutics and diagnostics. Even a gene of unknown function may have medical utility if its expression pattern can be determined.

To achieve this goal, methods and technologies operating reliably with many samples in high-throughput and in parallel are major requirements. The human genome is sequenced, but only a minority of genes has been assigned a function. Automated technology allows for high-throughput, resulting in fingerprints of diseased versus normal or developmentally distinct tissues. Differential gene expression can be most efficiently monitored by DNA hybridisation on arrays of oligonucleotides or cDNA clones. Having started from high-density filter membranes, cDNA microarrays nowadays are mainly used in chip or microscope slide format. In the past our group has shown that the same cDNA libraries used in gene expression analysis can be used for high-throughput protein expression and antibody screening on high-density filters and microarrays. Most importantly, these libraries connect recombinant proteins to clones identified by DNA hybridisation or sequencing, hence creating a direct link between gene catalogues and functional catalogues. Microarrays can now be used to go from an individual clone to a specific gene and its protein product or vice versa. Clone libraries have become amenable to database integration including all steps from DNA sequencing to functional assays of proteins.

The medical application of this information is expected to lead to new generations of products in the diagnostics and therapeutics market. However, genes will only be useful for drug development and medical diagnostics if their functions are known. To tackle the current limitations in the medical use of genome information, "functional genomics" or "functional proteomics" are now under development as a new research and development area.

With the introduction of automated technologies in the field of molecular biology and, especially, microarray technology, genome and gene expression analysis have been accelerated enormously. Microarray technology was enabled by the development of devices that can array biological samples at high density and with high precision [1]. Oligonucleotide and cDNA microarrays have become hot commodities, representing thousands of individual genes arrayed on filter or glass slide supports (The chipping forecast, Nature Genetics supplement 1999 [2]). To examine variation in gene expression, sets of oligonucleotides or complex probes, generated by reverse transcription of RNA from different tissues and cell lines, are hybridised on the arrays [3]. cDNA microarrays have already been used to profile human tissues like bone marrow, brain, prostate and heart [4] and complex diseases such as rheumatoid arthritis [5] and cancer [6, 7]. However, the DNA chip technology is still hampered by the lack of common quality standards that enable the comparison of results obtained in different laboratories and with arrays of different origin [8, 9]. Nonetheless, protein chips are already emerging to follow DNA chips as tools for automated and miniaturised functional analysis [10, 11]. Analogous to DNA microarrays, protein arrays offer the opportunity to screen thousands of immobilised biomolecules at a time, using steadily reduced amounts of sample.

2
From 2D Electrophoresis and Microtitre Plates to Microarrays of Biomolecules

Two-dimensional gel electrophoresis separates proteins according to size and charge, therefore allowing the study of cell, tissue and even whole organism proteomes [12]. Until recently, however, the identification of the thousands of separated proteins used to be a major challenge. With the introduction of new and automated mass spectrometric protein identification procedures, the high throughput identification of the separated proteins is much simplified [13] and allows us to generate catalogues of expressed proteins in a cell or tissue of interest. Nevertheless, as the separated proteins are obtained in denatured form and in limited amounts, the expression of a protein of interest in recombinant form is usually required for functional characterisation. The classical array format in proteomics, the microtitre plate, is a well established and still widely used standard in medical diagnostics. To increase the number of samples and decrease reagent volume, the 96-well microtitre plate has been developed further to plates with 384 and 1536 wells, maintaining the original plate footprint. As it happened in DNA analysis earlier, the microtitre plate is now gradually being replaced by microarrays on flat surfaces such as glass slides ("chips") or membranes.

The format and the preparation of protein microarrays depends on the nature of the immobilised biomolecule and its application. While peptide arrays are manufactured synthetically directly on the support [14], proteins are delivered using either pin-based spotting or liquid microdispensing. To date, the most commonly arrayed proteins are antibodies, since they are robust molecules which can be easily handled and immobilised by standard procedures without loss of activity.

3
Requirements for Protein Arrays

For protein arrays, resources of large numbers of proteins, preferably in purified form, represent a major technical challenge. While in prokaryotes genomic fragments can be directly cloned into expression vectors, intronic sequences prohibit this strategy in eukaryotes. In systems such as human tissues, full-length cDNA clones have to be isolated before protein expression can be started. High-throughput subcloning of open reading frames has been described [15] but remains a major difficulty if complete proteomes of higher organisms represent the region of interest. To overcome these problems, arrayed cDNA expression libraries, cloned in bacterial and yeast expression vectors, have been developed in our laboratory. These libraries are generated by standard DNA cloning procedures and characterised by oligonucleotide fingerprinting to be screened for the properties of their expression products. Furthermore, these libraries do represent an immortal source for large numbers of recombinant proteins [16–18]. In addition, expression libraries eliminate the need to construct individual expression systems for every protein of interest and, by arraying, the expression products of complete libraries can be characterised in parallel. On the other hand, a large proportion of clones do not express their insert in a suitable form, mainly due to cDNA fragments being fused to the vector-encoded start codon in the wrong reading frame. Therefore, non-expression clones have to be identified and removed from the library. To identify expression clones, hundreds of thousands of clones are arrayed on filter membranes and protein expression is induced. By detection of a His_6-tag peptide fused to the protein products, desired expression clones are identified and re-arrayed into a non redundant daughter expression library. The protein filter array technology was further developed to increase spot density and to facilitate the arraying of purified proteins. Lueking et al. have used automated arraying from liquid expression cultures using a pin-based, flat-bed gridding robot [19]. For this purpose, 96 proteins of the human foetal brain cDNA library hEx1 [17] were expressed in liquid bacterial cultures, and solutions were spotted onto polyvinylidene difluoride (PVDF) filters, either as crude lysates or after purification by Ni-NTA immobilised metal affinity chromatography (IMAC). In all 4,800 samples were placed onto polyacrylamide-coated microscopic slides and simultaneously screened, using a hybridisation automat, applying minimal amounts of reagents (less than 100 µL antibody solution; A. Lueking, personal communication). Sharp and well-localised signals allowed the detection of 250 attomol or 10 pg of a spotted test protein (GAPDH, glyceraldehyde-3-phosphate dehydrogenase, Swiss-Prot P04406).

4
Planar Immobilisation of Proteins

The Lehrach laboratory mainly uses solid pins routinely for spotting nanolitre volumes of proteins although proteins can be delivered onto solid supports by either split pin-based spotting or microdispensing devices. In contrast to other techniques, solid pins are less sensitive to variation of sample viscosity than slit pins or microdispensing systems and are much easier to clean to prevent any cross contaminations [19, 20]. A ring and pin arraying device was used by MacBeath and Schreiber to produce a microarray of 10,800 spots of two distinct proteins (protein G and an FKBP12 binding domain), which were then specifically detected with fluorescently labelled IgG and FKBP12, respectively. Although the detection scheme is this article is really clever, it remains unclear why the authors did array 10800 replicates made of only two proteins [21]. As an alternative to modified glass surfaces gel immobilisation matrices show high binding capacities and provide the proteins with a nearly (97% water in the buffer system) native environment. This is a key feature in protein array research and is important for the reactivity and specificity of the arrayed proteins. The technique of immobilisation is substantial both for effective concentration and orientation of immobilised proteins or antibodies on the surface. A variety of methods have been reported, including the adsorption to charged or hydrophobic surfaces, covalent cross-linking or specific binding via tags (e.g., His_6, biotin/avidin system).

The density of protein molecules immobilised on the support is mainly determined by the surface structure. A flat, two-dimensional surface offers less binding capacity than the three-dimensional structure of a filter membrane or a polyacrylamide gel layer. Mirzabekov and co-workers produced three-dimensional polyacrylamide gel pad microarrays providing a more than 100 times greater immobilisation capacity than two-dimensional glass supports, thus increasing the sensitivity of measurements considerably [22]. The gel pads are separated by a hydrophobic glass surface and provide a native, aqueous environment and can accommodate proteins of up to 400,000 Dalton in size [23]. Enzymatic activity of several enzymes like horseradish peroxidase, alkaline phosphatase and β-D-glucuronidase has been detected in these hydrogel pads. Prestructured surfaces consisting of hydrophilic spots on hydrophobic surfaces have also been reported for protein arraying [24, 25]. The hydrophobic surface prevents the aqueous drops applied to the hydrophilic spots from mixing and creates wall-minimised reaction vessels, where the interactions can be monitored in solution. In combination with state of the art microfabrication procedures, prestructured surfaces allow the introduction of three-dimensional microstructures on a chip, offering a number of additional options for experimentation like on-line monitoring of the interaction kinetics. Such microfluidic devices can be equipped with channels for transporting reagents to immobilised target molecules. At the present time microfluidic chips do certainly offer specific advantages over planar microarrays [26] but due to their complex production procedures and the high surface to volume ratio that represents a potential non-specific binding site for the analyte, their development and applications are still at an early stage.

5
Detection of Molecular Interactions on Microarrays

On DNA microarrays, hybridisation events are detected using fluorescently or radioactively labelled probe molecules [27]. A corresponding approach for the detection of protein-protein, protein-DNA and protein-small molecule interactions has been reported, the "universal protein array system" (UPA), consisting of filter membrane arrays of purified proteins [28]. Specific binding properties of the immobilised proteins on the low-density UPA arrays were demonstrated with various radiolabelled protein, DNA, RNA and small molecule ligands. By washing the membrane with different salt conditions, high-affinity protein-protein interactions could be distinguished. In addition to fluorescent dyes and radioisotopes, a wide range of detection options exists for protein and antibody arrays (reviewed in [29]). Unlabelled ligands can be identified indirectly by using a secondary antibody (sandwich assay). Alternatively to these non-competitive formats, various competitive assays, relying on competition of the ligand with labelled tracers, are in use. Protein chips for direct measurement of protein mass by matrix-assisted laser desorption-ionisation time-of-flight (MALDI-TOF) mass spectrometry have been described [20, 24, 30].

6
Applications of Protein Arrays

A large variety of assays has been adapted to utilise protein microarrays. At its current state, the detection of immobilised antigens with antibodies is still the most common application. Protein and antibody arrays have been used for the selection and characterisation of novel antibodies from phage display libraries and for the identification of antigens (e. g., involved in autoimmune diseases).

Phage display antibody libraries have been developed for the in vitro selection of antibodies as an alternative to animal immunisation (reviewed in [32], [53]). For this purpose, recombinant immunoglobulin gene libraries are cloned in phagemid vectors and antibody fragments are displayed as fusion proteins on the surface of bacteriophage (reviewed in [34]). Recently, protein arrays of the cDNA expression library hEx1 [17] were used to identify antigens recognised by randomly selected antibody fragments from a phage display antibody library [35]. Screening 12 different antibody fragments on an array of 27,000 expression clones, delivered four novel and highly specific antigen-antibody pairs. In a related approach, antibody arrays were used for the identification of specific antibody-producing bacteria [36]. For this purpose, bacteria containing phagemid selected from a phage antibody library by in vitro panning on chosen antigens were arrayed on filter membranes. After cell growth, antibody production was induced and specifically binding antibodies were captured and identified on a second, antigen-coated membrane. By screening 18,342 antibody clones at a time, highly specific antibodies were selected after just one round of panning.

In autoimmune diseases self-reacting antibodies, i. e., produced against the organism's own proteins and epitopes, play an important role in the clinical manifestation of the diseases. Therefore, the generation of an antibody profile of patients with autoimmune disease is believed to be medically relevant and informative. Characterisation of autoimmune patient sera on protein chips would allow the diagnosis of autoimmune diseases based upon the presence of specific auto-antibodies. So far, for the identification of antigens recognised by auto-antibodies, sera were hybridised to uncharacterised gt11 cDNA phage libraries or to tissue extracts separated by 1D or 2D gel electrophoresis [37, 38]. The following characterisation of the identified antigens is labour intensive, also requiring expensive sequencing of the found proteins. Such characterisation resulted in novel functions attributed to these proteins, which can then be used as potential therapeutical targets [39]. To simplify the characterisation of auto-antibodies, serum can be applied to protein arrays containing large numbers of recombinant proteins of known identity. Moreover, using protein arrays will overcome the problems associated with protein level variation in natural tissue extracts and hence increase reproducibility. The application of protein chips allows us to determine the binding profile of the autoimmune antibodies of each patient and for each disease. Once disease-specific antigens are known, it is possible to create a diagnostic protein array. As shown by Lueking et al., apparently specific monoclonal antibodies (α-HSP90, α-β-tubulin) showed considerable cross-reactivity with other proteins following incubation on protein microarrays, consisting of 96 in liquid bacterial cultures expressed proteins of the hEx1 library [19]. In a way, this is not surprising, as antibodies are not usually tested against whole libraries of proteins. However, in immunohistochemical or physiological studies against whole cells or tissue extracts, this cross-reactivity of antibodies can lead to false interpretations. Therefore, the characterisation of the binding specificity of antibodies used extensively in diagnostic tools is of prime importance.

7
Outlook

To achieve standardised microarrays carrying thousands of verified recombinant proteins, high-throughput methods for protein expression and purification are required. This has to be accompanied with a pipeline for the identification and verification of expression products. Initial experiments have shown that especially MALDI-MS is a powerful tool to monitor the quality of recombinant proteins. Combining protein expression and purification in array (microtitre plate) format with high-throughput protein mass determination by mass spectrometry leads to a large number of identified library clones and their corresponding expression products [17, 20]. This approach can be used either to identify unknown clones from expression libraries or to verify expression products generated at high-throughput. In addition mass spectrometric data allows us to compare recorded spectra from recombinant and native proteins, which results in unambiguous protein identification in, e. g., 2D protein gels.

A very crucial part in protein array technology is played by the deposition of the proteins on a suitable surface. As described earlier, the nature of the surface with influencing parameters like charge, viscosity, pore size, pH, binding capacity, unspecific protein binding, etc. is essential for the generation of protein arrays that contain the proteins in an biologically active shape and form. New surfaces will lead to more native "living protein arrays". The basics for such a technology has been described by the analysis of protein-protein interactions in *S. cerevisiae* using the yeast two-hybrid screens in array format [11,31]. So called "living arrays" were constructed consisting of a nearly complete set of yeast open reading frames cloned as fusions with the Gal4 activation domain. This clone set was co-transformed with a set of putative interaction partners cloned as fusions to the Gal4 DNA binding domain and subsequently arrayed on filter membranes. Protein-protein interaction was detected by arraying of the co-transformed clone set on selective media. By screening 5,345 yeast open reading frame-Gal4 activation domain fusions with 195 Gal4 DNA binding domain fusions, 957 putative interactions, involving 1,004 yeast proteins, were identified. The recombinant expression of all open reading frames of *Saccharomyces cerevisiae* has already been achieved. A nearly complete collection of yeast strains for the expression of 6,144 open reading frames as fusion proteins was generated, divided in pools and screened for biological activities. Collections like this can form the basis of future protein microarrays by representing a large portion of gene products. As an example for an functional assay, 119 protein kinases were expressed, purified as GST fusion proteins, arrayed and cross-linked in a protein chip format and assayed for autophosphorylation by treatment with radiolabelled ATP. Also, substrate specificity was assayed with protein chips each carrying one of a set of kinase substrates. The kinases and the radiolabelled ATP were arrayed by pipetting onto the substrate coated surfaces and phosphorylation was monitored [40–42].

In addition to all array applications, clones and their recombinant proteins will form the basis for "structural genomics", a research field that aims to resolve the molecular structure of biomolecules and biomolecule complexes. Within the Berlin Protein Structure factory the same proteins that are deposited onto arrays are recombinantly expressed in yeast and *E. coli* and are subjects for crystallisation and the subsequent X-ray scattering as well as for NMR experiments. Similar technology to the technology described here is used with the immortal protein resources to do the probe preparation for structural genomics. One example is a high-precision sub-microlitre liquid dispensing system that has been developed for the preparation of hanging drop arrays for protein crystallisation. These arrays consist of 2 μL to 100 nL drops and are used to screen for suitable buffer and salt conditions for protein crystallisation [25]. The technology developed for systematic data generation in genomics and proteomics will enable us to crystallise proteins that are available only in very tiny amounts and, therefore, contribute to our understanding of genome structure and physiological function.

8
References

1. Maier E, Meier-Ewert S, Bancroft D, et al (1997) Automated array technologies for gene expression profiling. Drug Discovery Today 2(8):315–324
2. Lander ES (1999) Array of hope. Nat Genet 21(1 Suppl):3–4
3. Meier-Ewert S, Lange J, Gerst H, et al (1998) Comparative gene expression profiling by oligonucleotide fingerprinting. Nucleic Acids Res 26(9):2216–2223
4. Schena M, Shalon D, Heller R, et al (1996) Parallel human genome analysis: microarray-based expression monitoring of 1000 genes. Proc Natl Acad Sci USA 93(20):10614–10619

5. Heller RA, Schena M, Chai A, et al (1997) Discovery and analysis of inflammatory disease-related genes using cDNA microarrays. Proc Natl Acad Sci USA 94(6):2150–2155
6. Khan J, Saal LH, Bittner ML, et al (1999) Expression profiling in cancer using cDNA microarrays. Electrophoresis 20(2):223–229
7. Khan J, Bittner ML, Chen Y, et al (1999) DNA microarray technology: the anticipated impact on the study of human disease. Biochim Biophys Acta 1423(2):M17–28
8. Eickhoff H, Schuchhardt J, Ivanov I, et al (2000) Tissue gene expression profiling using arrayed normalized cDNA libraries. Genome Res 10(8):1230–1240
9. Schuchhardt J, Beule D, Malik A, et al (2000) Normalization strategies for cDNA microarrays. Nucleic Acids Res 28(10):E47
10. Walter G, Büssow K, Cahill D, et al (2000) Protein arrays for gene expression and molecular interaction screening. Curr Opin Microbiol 3(3):298–302
11. Emili AQ, Cagney G (2000) Large-scale functional analysis using peptide or protein arrays. Nat Biotechnol 18(4):393–397
12. Klose J, Kobalz U (1995) 2-dimensional electrophoresis of proteins – an updated protocol and implications for a functional-analysis of the genome. Electrophoresis 16(N6):1034–1059
13. Gauss C, Kalkum M, Lowe M, et al (1999) Analysis of the mouse proteome. (I) Brain proteins: separation by two-dimensional electrophoresis and identification by mass spectrometry and genetic variation. Electrophoresis 20(3):575–600
14. Frank R (1992) Spot synthesis: an easy technique for the positionally addressable parallel chemical synthesis on a membrane support. Tetrahedron 48:9217–9232
15. Heyman JA, Cornthwaite J, Foncerrada L, et al (1999) Genome-scale cloning and expression of individual open reading frames using topoisomerase I-mediated ligation. Genome Res 9(4):383–392
16. Büssow K, Cahill D, Nietfeld W, et al (1998) A method for global protein expression and antibody screening on high-density filters of an arrayed cDNA library. Nucleic Acids Res 26(21):5007–5008
17. Büssow K, Nordhoff E, Lübbert C, et al (2000) A human cDNA library for high-throughput protein expression screening. Genomics 65(1):1–8
18. Lueking A, Holz C, Gothold C, et al (2000) A system for dual protein expression in *Pichia pastoris* and *Escherichia coli*. Protein Expr Purif 20:372–378
19. Lueking A, Horn M, Eickhoff H, et al (1999)Protein microarrays for gene expression and antibody screening. Anal Biochem 270(1):103–111
20. Egelhofer V, Büssow K, Luebbert C, et al (2000) Improvements in Protein Identification by MALDI-TOF-MS Peptide Mapping. Anal Chem 72(13):2741–2750
21. MacBeath G, Schreiber SL (2000) Printing proteins as microarrays for high-throughput function determination. Science 289(5485):1760–1763
22. Parinov S, Barsky V, Yershov G, et al (1996) DNA sequencing by hybridization to microchip octa-and decanucleotides extended by stacked pentanucleotides. Nucleic Acids Res 24(15):2998–3004
23. Arenkov P, Kukhtin A, Gemmell A, et al (2000) Protein microchips: use for immunoassay and enzymatic reactions. Anal Biochem 278(2):123–131
24. Schuerenberg M, Luebbert C, Eickhoff H, et al (2000) Prestructured MALDI-MS sample supports. Anal Chem 72(15):3436–3442
25. Müller U, Nyarsik L, Horn M, et al (2001) Development of a technology for automation and miniaturisation of protein crystallisation. J Biotechnol (in press)
26. Sanders GHW, Manz A (2000) Chip-based microsystems for genomic and proteomic analysis. Trends Anal Chem 19(6):364–378
27. Cheung VG, Morley M, Aguilar F, et al (1999) Making and reading microarrays. Nat Genet 21(1 Suppl):15–19
28. Ge H (2000) UPA, a universal protein array system for quantitative detection of protein-protein, protein-DNA, protein-RNA and protein-ligand interactions. Nucleic Acids Res 28(2):e3
29. Rogers KR (2000) Principles of affinity-based biosensors. Mol Biotechnol 14(2):109–129

30. Merchant M, Weinberger SR (2000) Recent advancements in surface-enhanced laser desorption/ionization-time of flight-mass spectrometry. Electrophoresis 21(6):1164–1177
31. Uetz P, Giot L, Cagney G, et al (2000) A comprehensive analysis of protein-protein interactions in *Saccharomyces cerevisiae*. Nature 403(6770):623–627
32. Holt LJ, Enever C, de Wildt RM, et al (2000) The use of recombinant antibodies in proteomics. Curr Opin Biotechnol 11(5):445–449
33. Hoogenboom HR, de Bruïne AP, Hufton SE, et al (1998) Antibody phage display technology and its applications. Immunotechnology 4(1):1–20
34. Collins J (1997) Phage Display. In: Moos WH, Pavia MR, Ellington A, Kay BK (eds) Annual Reports in Combinatorial Chemistry and Molecular Diversity, vol 1. Kluwer Academic Pub, pp 210–262
35. Holt LJ, Büssow K, Walter G, et al (2000) By-passing selection: direct screening for antibody-antigen interactions using protein arrays. Nucleic Acids Res 28(15):E72
36. de Wildt RM, Mundy CR, Gorick BD, et al (2000) Antibody arrays for high-throughput screening of antibody-antigen interactions. Nat Biotechnol 18(9):989–994
37. Latif N, Baker CS, Dunn MJ, et al (1993) Frequency and specificity of antiheart antibodies in patients with dilated cardiomyopathy detected using SDS-PAGE and western blotting. J Am Coll Cardiol 22(5):1378–1384
38. Pohlner K, Portig I, Pankuweit S, et al (1997) Identification of mitochondrial antigens recognized by antibodies in sera of patients with idiopathic dilated cardiomyopathy by two-dimensional gel electrophoresis and protein sequencing. Am J Cardiol 80(8):1040–1045
39. McCurdy DK, Tai LQ, Nguyen J, et al (1998) MAGE Xp-2: a member of the MAGE gene family isolated from an expression library using systemic lupus erythematosus sera. Mol Genet Metab 63(1):3–13
40. Joos TO, Schrenk M, Hopfl P, et al (2000) A microarray enzyme-linked immunosorbent assay for autoimmune diagnostics. Electrophoresis 21(13):2641–2650
41. Martzen MR, McCraith SM, Spinelli SL, et al (1999) A biochemical genomics approach for identifying genes by the activity of their products. Science 286(5442):1153–1155
42. Cohen CB, Chin-Dixon E, Jeong S, et al (1999) A microchip-based enzyme assay for protein kinase A. Anal Biochem 273(1):89–97

Received: June 2001

Microarray Data Representation, Annotation and Storage

Alvis Brazma[1] · Ugis Sarkans[1] · Alan Robinson[1] · Jaak Vilo[1] · Martin Vingron[2] · Jörg Hoheisel[3] · Kurt Fellenberg[3]

[1] EMBL Outstation – Hinxton, European Bioinformatics Institute, Wellcome Trust Genome Campus, Hinxton, Cambridge CB10 1SD, UK. *E-mail: brazma@ebi.ac.uk*
[2] Max-Planck-Institut für Molekulare Genetik, Ihnestrasse 73, 14195 Berlin, Germany
[3] Functional Genome Analysis, Deutsches Krebsforschungszentrum, Im Neuenheimer Feld 506, 69120 Heidelberg, Germany

Management and analysis of the huge amounts of data produced by microarray experiments is becoming one of the major bottlenecks in the utilization of this high-throughput technology. We describe the basic design of a microarray gene expression database to help microarray users and their informatics teams to set up their information services. We describe two data models – a simpler one called ArrayExpressB and the complete model ArrayExpressC, and discuss some implementation issues. For latest developments see http: www.ebi.ac.uk/arrayexpress.

Keywords: Microarrays, Bioinformatics, Databases, Gene expression, Object modeling, UML

1	Introduction	115
2	Minimum Information About a Microarray Experiment	116
3	Structure and Design of the ArrayExpress Database	117
4	Object Modeling, UML and OMT Notations	122
5	Basic Model – ArrayExpressB	123
5.1	Experiment as a Whole	123
5.1.1	Experiment	123
5.2	Hybridization	123
5.3	Describing Arrays and Array Types	124
5.3.1	Array Type	124
5.3.2	Element	124
5.3.3	Biosequence	125
5.4	Extract Preparation	125
5.4.1	SampleSource	125
5.4.2	Sample	126
5.4.3	Treatment	126
5.4.4	Extraction and Extract	126
5.4.5	Labeling and LabeledExtract	127
5.5	Array Scanning and Analysis	127
5.5.1	Image	127

Advances in Biochemical Engineering/
Biotechnology, Vol. 77
Managing Editor: T. Scheper
© Springer-Verlag Berlin Heidelberg 2002

5.5.2   ImageAnalysis 127
5.5.3   ExpressionValueSet, ExpressionValue and ExpressionValueType . 128
5.6   Common Constructs 128
5.6.1   Description 129
5.6.2   Protocol 129
5.6.3   Property 129
5.6.4   PropertyType 129
5.6.5   ExternalReference 129
5.6.6   LiteratureReference 130
5.6.7   DatabaseReference 130
5.6.8   OntologyReference 130
5.6.9   WebReference 130

6   **Complete Model – ArrayExpressC** 130

6.1   Experiment as a Whole 131
6.2   Hybridization 131
6.3   Describing Arrays and Array Types 131
6.3.1   Array and ArrayBatch 131
6.3.2   Element 131
6.3.3   Reporter 132
6.4   Extract Preparation 132
6.4.1   Biomaterial 132
6.4.2   Event 133
6.4.3   Sample 133
6.5   Array Scanning and Analysis 133
6.5.1   ImageAnalysis 134
6.5.2   ExpressionValueSet 134
6.5.3   ExpressionValue 134
6.5.4   ExpressionValueDimension 134
6.5.5   ExpressionValueType 135
6.5.6   CompositeElement 135
6.5.7   CompositeSample 135
6.5.8   Transformation 136
6.6   Common Constructs 136

7   **Implementation Issues** 136

7.1   Database Implementation 136
7.2   Data Queries, Mining and Visualization 137

8   **Discussion** 137

9   **References** 139

1
Introduction

Management and analysis of the huge amounts of data produced by microarray experiments is becoming one of the major bottlenecks in the utilization of this high-throughput technology. Organizing and storing microarray data so that they can be queried and analyzed is not a trivial problem for several reasons.

First of all microarray data have meaning only in the context of the particular biological sample and exact experimental conditions under which the experiments have been performed. Developing ways to describe and represent biological conditions in a precise way that can be queried is one of the major research areas in bioinformatics.

The second major problem originates from the lack of standards in microarray technologies and data analysis methods. The raw data produced by array technologies are images, while, conceptually, gene expression data is a table representing expression levels of particular genes in particular samples (for instance we could assume that each row in the table represents a gene, and each column a sample – we call such a table a *gene expression matrix*, while the values in the matrix are often referred to as the *gene expression levels*). Ideally, it would be desirable to measure the expression levels in some natural units, such as the number of mRNA molecules per cell, and also to have an error-estimate associated with each value. Unfortunately, the accuracy of most measurement procedures used at present is not sufficient to reliably assess the abundance of gene products in such objective units. Moreover, the relationship between the raw images and the gene expression matrix is not straightforward. To obtain any meaningful gene expression data from such images, they have to be quantitized and data from several images have to be combined and normalized, and the results recorded in the database.

In the absence of standard controls, the gene expression data from different sources effectively use different measurement units; the relative values of which are typically unknown and may even vary depending on the measurement range. This makes it necessary to store in the database not only the gene expression data matrix, but also the raw and intermediate data, and a detailed description as to how the processed values are obtained. This complicates the database design enormously. The size of the datasets (that can easily reach terabytes) adds to the difficulties.

In this paper we describe the basic design of a microarray gene expression database to help microarray users and their informatics teams set up their information services. The first version of the microarray database object model *ArrayExpress*, which is described here, was developed at the European Bioinformatics Institute (EBI) in collaboration with the German Cancer Research Centre (DKFZ) and posted on the Internet in November, 1999 (see (http://www.ebi.ac.uk/arrayexpress)). A relational database implementation *maxd* based on the ArrayExpress object model was conducted at the University of Manchester (http://bioinf.man.ac.uk/microarray/resources.html) and is widely used for microarray laboratory informatics support.

The first ArrayExpress version was limited in several respects. Firstly, it was essentially array element (spot) centric, rather than gene centric. Concretely, the model assumed that each gene is represented by one spot on the array. It also did not allow information from replicate experiments to be easily summarized. To deal with these and other limitations we have developed a revised version of ArrayExpress, which we describe here. In fact we will describe two models – a simpler one called the Basic model (*ArrayExpressB*), and a more elaborate one, called the Complete model (*ArrayExpressC*). The new ArrayExpress model has been considerably influenced by the object model used for the *Genomics Knowledge Platform* developed by Incyte Genomics.

A standard way to describe biological samples and conditions is needed if the expression data are to be queried and compared. This requires the use of standard ontologies and controlled vocabularies for biological sample description, which at the moment are only under development. The Microarray Gene Expression Database (MGED) group was founded in 1999 to facilitate the adoption of standards for DNA-array experiment annotation and data representation, as well as the introduction of standard experimental controls and data normalization methods. The MGED group has recently completed their first set of recommendations (see (http://www.mged.org)). At the basis of these recommendations is the formulation of the minimum information about a microarray experiment that researchers are encouraged to report in order to ensure the interpretability of the experimental results, as well as their potential verification by third parties. The recommendations are briefly described in the following sections. The database design described here can be used to capture information complying with these MGED recommendations.

2
Minimum Information About a Microarray Experiment

It is important that all the information that may be needed to interpret the results of a microarray experiment at a later date, and possibly to verify the experiment, is captured and recorded in the database. The microarray data annotations working-group of MGED has developed a list of requirements that have to be satisfied to meet these goals, called the "Minimum Information About a Microarray Experiment" (MIAME).

There are two simple principles underpinning MIAME. First, microarray data should be annotated in sufficient detail to be of most use to third parties. Second, while the microarray technology is still rapidly developing, it would be counterproductive to try to impose on researchers the use of any particular platforms or software, and any particular methods of data analysis. Instead, the standards should simply require revealing of data in sufficient detail.

The minimum information about a published microarray based gene expression experiment includes the following six parts:

1. experimental design: the set of the hybridization experiments as a whole;
2. array design: each array used and each element (spot) on the array;
3. samples: samples used, the extract preparation and labeling;

4. hybridizations: procedures and parameters;
5. measurements: images, quantitation, specifications; and
6. controls: types, values, specifications;
   Details of 1 – 6 can be found on (http://www.mged.org).

MIAME is aimed at cooperative data providers, and not to close all possible loopholes in providing the crucial information. The goal is to produce a workable standard that neither imposes an undue burden on scientists nor limits their ability to communicate their unique results. Among the concepts in the standard are sets of 'qualifier, value, source' triplets, where 'source' is either user defined, or a reference to an externally defined ontology or controlled vocabulary, such as the species taxonomy database at NCBI (http://www.ncbi.nlm.nih.gov/Taxonomy). Where necessary, authors are encouraged to define their own qualifiers and provide the appropriate values so that the list as a whole gives sufficient information to interpret the particular part of the experiment. Judgment regarding the necessary level of detail is left to the data providers. In future, these 'voluntary' qualifier lists may be gradually substituted by required fields, as the respective ontologies are developed. Parts of the MIAME can be provided as a reference or link to an externally existing description. For instance, for commercial or other standard arrays, all the required information should be normally described only once by the array provider and referenced by the users. Standard protocols should also normally be stored only once. It is necessary that either a valid reference or the information itself is provided.

We recommend that all the information about microarray experiments specified in MIAME is captured by the users in a database. The database design described below can be used to satisfy the MIAME requirements.

3
Structure and Design of the ArrayExpress Database

The presented microarray database design allows the capture of information about all the experimental steps that are performed to measure gene expression: preparing the samples, extracting and labeling gene transcripts from each sample, hybridization of these extracts onto the array, scanning the hybridized array, analyzing the scanned image to identify and quantify each element on the array, and deriving the gene expression levels from this quantification. The last step may combine values from several elements related to the same gene (possibly from different hybridizations).

The basic data unit that can be stored in the database is called *Experiment*. An *Experiment* consists of one or more *Hybridizations* between labeled extracts obtained from *Samples* and reporters that are printed on the *Array*. The hybridized array is analyzed by applying scanning and image analysis protocols, and an *ExpressionValue* is attributed to each spot on the array. The overall structure of the database, its top-level objects and the relationships between them are given in Fig. 1.

One of the considerations behind the proposed design is that the model should be general enough to capture information from all existing microarray plat-

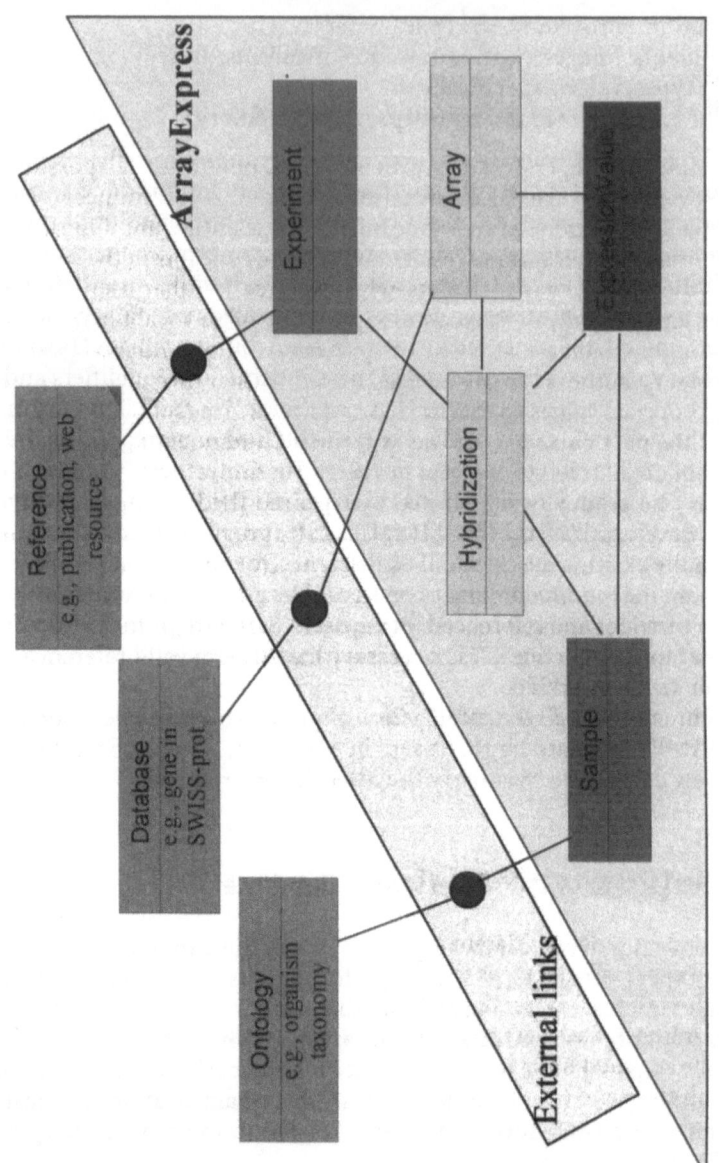

Fig. 1. Overall structure of ArrayExpress

forms, and even some possible future platforms. For laboratories using only one particular array platform, such a generality may not be necessary, but, on the other hand, it would enable them to switch to new platforms more easily.

The second aspect of the ArrayExpress model, which distinguishes it from similar simpler designs, is the ability to capture raw and intermediate data originating from microarray experiments. For instance, many of the existing public datasets report only the background-subtracted signal ratios for the popular two channel (Cy3/Cy5) technologies [1], which result in the loss of all the quality information characterizing each individual data-point. ArrayExpress allows the capture of complete outputs from image quantitation software regardless of the particular image analysis algorithm or software package used. Quantities such as the standard deviation of pixel intensities within a spot or number of pixels in a spot provide useful information about the quality of the particular data-point, which is important in many applications. Laboratories using a particular scanner and image analysis software may simplify the design, but the tradeoff between the simplicity and the flexibility to update the technology without changing the database has to be kept in mind.

It should also be noted that we cannot assume that an array element (spot) and a gene have a one-to-one relationship between them – in fact, for oligonucleotide arrays, there may be far too many relationships between genes and elements on the array (for instance, if there is a need to distinguish between closely homologous genes). A one gene, many spots relationship arises also when dealing with replicate experiments. The ArrayExpress model allows the storage of the initial array element quantitations along with all the derived values which can be obtained by combining several primary element quantitations. The final derivations would normally reflect gene expression levels, either absolute or relative, depending on the particular technology.

Because of these complexities, we will start by describing a simplified Array-Express model – ArrayExpressB, after which we will add the details of the ArrayExpressC model. Some users may not need to implement the ArrayExpressC model for their particular purposes. We encourage users to assess whether the ArrayExpressB model is sufficient or not (data structured according to the ArrayExpressB model can be later imported into the ArrayExpressC model compliant database). Also, a whole range of "intermediate" models between the two extremes is possible. In fact, these models should be critically assessed by users and adjusted to the particular needs of their laboratories.

The ArrayExpressB and ArrayExpressC models are presented on Figs. 2 and 3, respectively. Models are described as Unified Modeling Language (UML) diagrams; however, so-called Object Modeling Technique (OMT) notation is used to represent association cardinalities, giving more compact representation. In the following sections we will describe the most essential feature of the ArrayExpress model, the full documentation is available on (http://www.ebi.ac.uk/arrayexpress/).

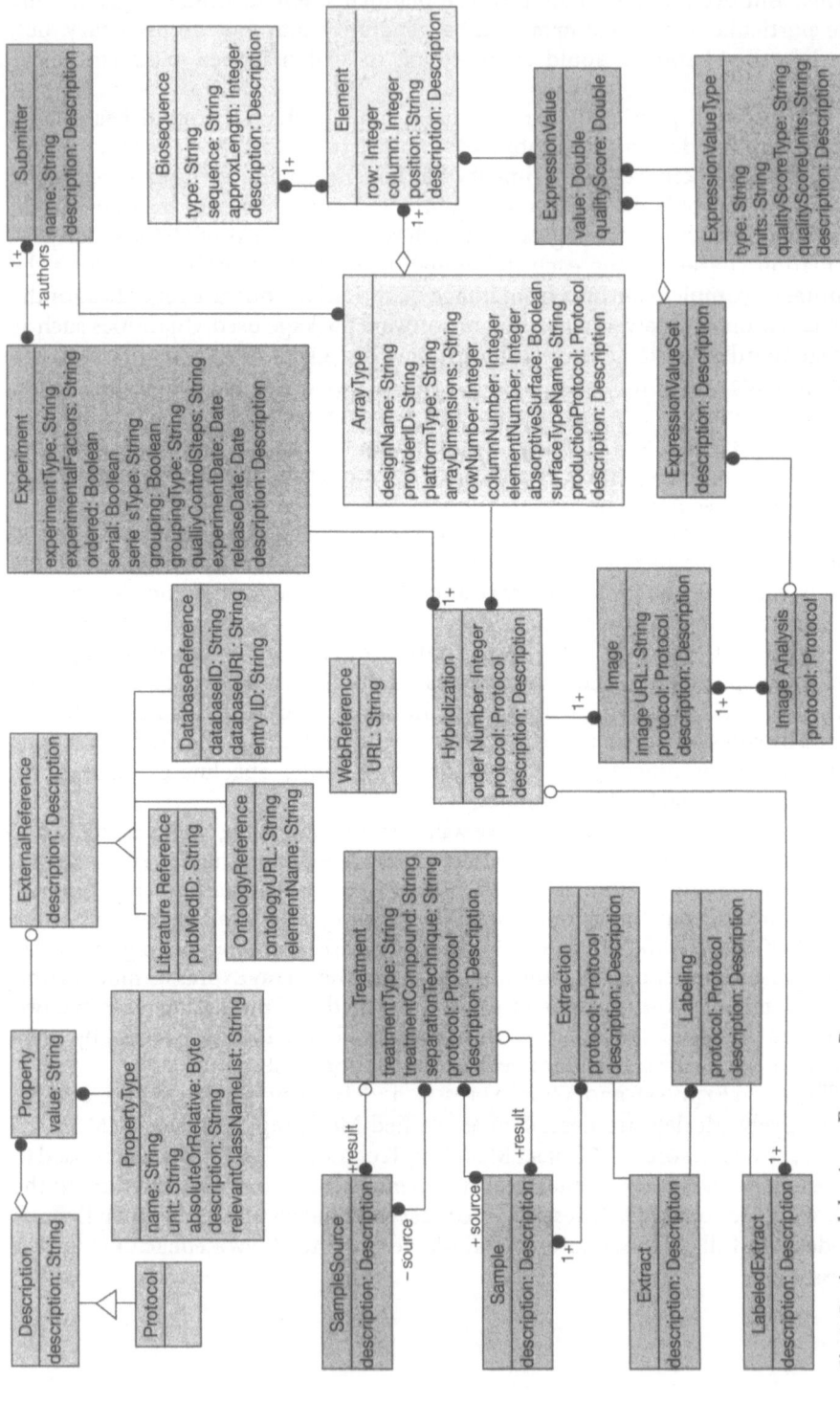

Fig. 2 Basic model – ArrayExpress B

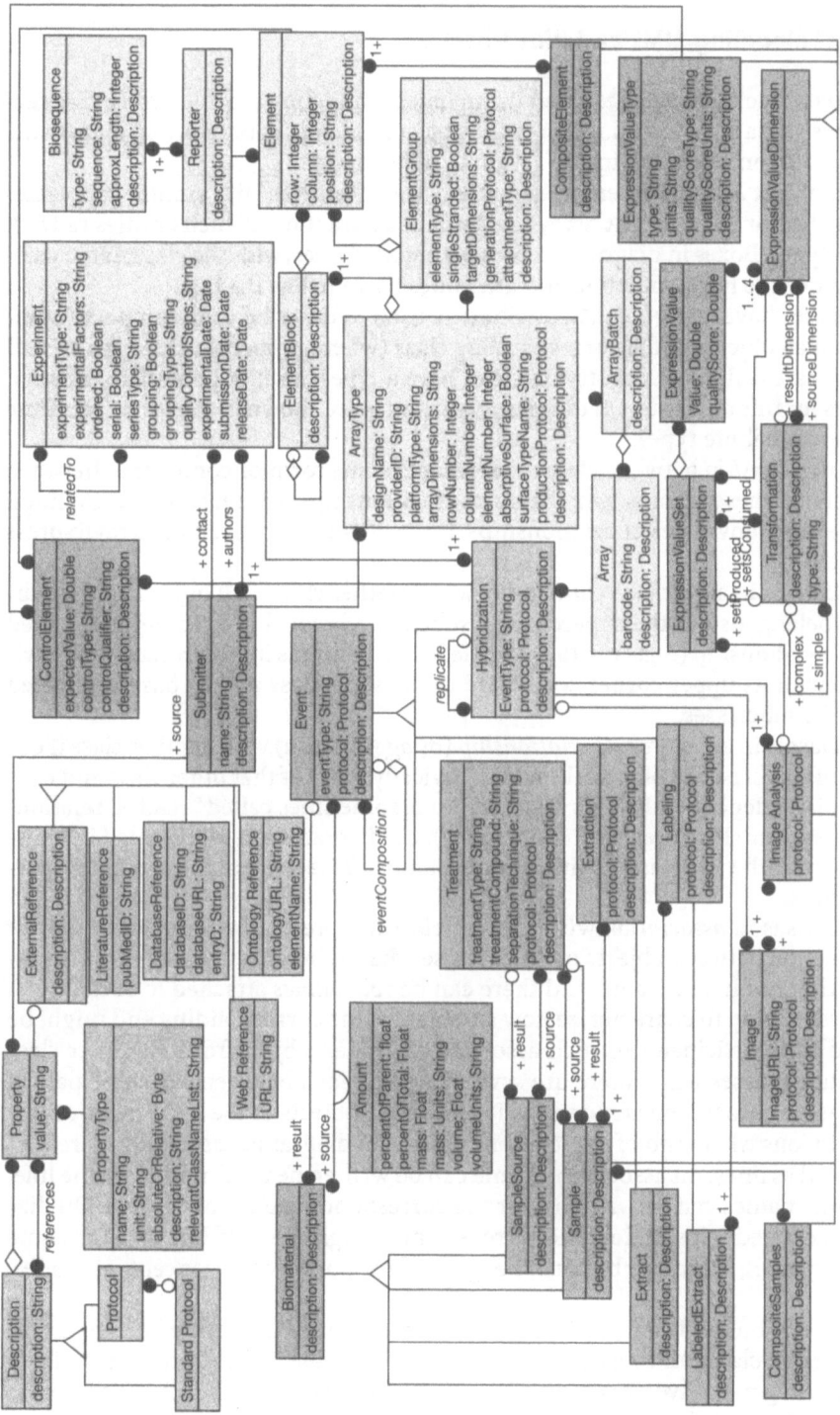

Fig. 3 Complete model – ArrayExpress C

4
Object Modeling, UML and OMT Notations

The database has been designed using object-modeling techniques. In this section we give a brief introduction to the object modeling approach and explain the used notation. For a full introduction to object modeling, see [2].

An object *class* represents a set of real-world entities with similar properties and behavior (e. g., a DNA array) or domain abstraction of such entities (a DNA array type). Boxes in diagrams represent object classes, with the class name written above the horizontal line and attributes listed below the line.

Classes have *attributes*. An attribute is used to describe some property common to all objects of the corresponding class (where property values can be and usually are different). An attribute can have a *type* – a set of potentially allowed values for this attribute. On diagrams, attributes are shown in the form ⟨attribute name⟩:⟨attribute type⟩.

A *relationship* between classes represents some form of connection between objects of these classes. On diagrams, relationships are represented by lines. There are many types of relationships, three of which are used in ArrayExpress models.

A class is in a *subclass relationship* with another class if all objects of the subclass belong also to the superclass. A subclass relationship is sometimes called "IS A" relationship (e. g., "car IS A vehicle"). On diagrams it is represented by a triangle with its upper corner connected to the superclass and its base connected to all its subclasses.

A class is in an *aggregate relationship* (or *aggregation*) with another class if objects of this class are physically and/or logically parts of that other class and cannot exist independently. This relationship is sometimes called "HAS A" relationship (e. g., car HAS A registration number). On diagrams it is represented by a line connecting the two classes, with a small diamond at the end of the container (parent) class.

A class is in *association* with another class if there can be some other type of connection between objects of these classes than subclass or aggregation. An association can have a name, and there can be *role names* attached to each end of the association to characterize how an object at the corresponding end might be called if it participates in such association with an object from the other class (useful for cases when there are several associations between the same pair of classes, an association is recursive, i. e., between objects of the same class, or it is not obvious what kind of association this is). On diagrams, associations are represented as lines. An association name can be written near the middle of the line, and role names can be written near the corresponding ends of the line. On diagrams created with Rational Rose role names might be prefixed by + or some other symbol; these symbols can be ignored for purposes of conceptual modeling.

Associations and aggregations can have *cardinalities*, indicating how many objects of one class can be associated with the same object of the other class. Each relationship may have two cardinalities, one for each end. Typical cardinalities are 1 (each object at the other end must have exactly one object at this end associated through this relationship), 0...1 (... can have no more than one ...), 0...* (.... can have any number ...), 1...* (... must have at least one ...). On diagrams, cardinalities are written near the corresponding relationship end. In OMT notation, which is more compact, relationship ends are transformed. No transformation means cardinality 1, small hollow circle means 0...1, small filled circle means 0...*, and small filled circle with "1+" written nearby means 1...*.

5
Basic Model – ArrayExpressB

In this section we describe the most essential elements of the ArrayExpress model, particularly the ones that are important for laboratory databases. The Complete ArrayExpress model will be described in the next section.

5.1
Experiment as a Whole

5.1.1
Experiment

This is the top-level class of ArrayExpress. An object of this class represents a description of a microarray experiment as a unit. Data is entered into the database by filling in one or more **Experiment** class objects together with all related objects.

An **Experiment** consists of one or more **Hybridizations**. **Experiment** class attributes are used to describe properties of a microarray experiment as a whole. They include:

– Experiment type, e.g., toxicity, drug efficacy, differential expression, normal vs diseased comparison, treated vs untreated comparison, time course, dose response, effect of gene knock-out, effect of gene knock-in, shock. Multiple types are possible.
– Experimental factors, e.g., time, dosage, severity, location, genetic background.
– A flag indicating whether **Hybridizations** of this **Experiment** are ordered or not; actual order of **Hybridizations** is determined by **orderNumber** attribute of the **Hybridization** class.

Experiment class has an attribute **description** which can hold a reference to a **Description** class object (described below), specifying publications in which this **Experiment** is documented, online supplementary resources and other relevant information which does not fit into provided attributes.

5.2
Hybridization

An object of the **Hybridization** class represents an instance of array hybridization, where a labeled extract from the target sample is hybridized with an array of a given array type, under a given set of conditions. A **Hybridization** is a part of an **Experiment**. One or more **LabeledExtracts** are used for a **Hybridization**, typically using differently labeled extracts. **Hybridization** results are scanned, resulting in one or more **Images**.

A **Hybridization** class attribute **orderNumber** can be used to record ordering of **Hybridizations** within an **Experiment** (if the order is relevant). The protocol used to hybridize the **LabeledExtract** to the array should be described using **protocol** attribute. Information that can be recorded here includes: solution characteristics (e.g., concentration of solutes), blocking agent, wash procedure, quantity of labeled target used, time, volume, temperature, and description of the hybridization instruments.

5.3
Describing Arrays and Array Types

5.3.1
ArrayType

ArrayType class objects correspond to microarray designs, not individual arrays. We expect that for standard arrays this information will be provided by the array producers, either commercial or resource centers. The same **ArrayType** can be used for many **Hybridizations**; therefore, a single **ArrayType** object can be associated with multiple **Hybridization** objects. Attributes of the **ArrayType** class include the following attributes:

– Array design name (e.g., "Stanford Human 10K set").
– Platform type, e.g., spotted or synthesized; cDNA, oligos or PCR products; plasmids or colonies.
– Array size, the number of rows and columns in the grid of **Elements**, or the number of elements if the layout is more complex than orthogonal rows and columns.
– Array substrate: absorptive or nonabsorptive, type name (e.g., glass, nylon).
– Production protocol.

Laboratories making their own arrays should normally capture this information by their laboratory information management systems (LIMS), which should ideally upload this information directly into the database.

5.3.2
Element

An **Element** represents a single element (spot) on the array or, more precisely, design of a single element as a part of the design of related **ArrayType**. Elements contain reporters used to detect the target molecules. A reporter is a set of identical (in principle) DNA molecules. For an **Element** its row and column (or, if layout is more complex – its position) can be described using **Element** class attributes. Other relevant information, e.g., production protocol or size, can be captured by the **description** attribute. Each **Element** object is related to one or more **Biosequence** objects.

5.3.3
Biosequence

Biosequence objects characterize the reporter contained within the respective **El-ement**. This information – what is the sequence of **Element's** reporter, what gene this reporter is thought to represent – can be regarded as an attribute of **Element** class. However, more complex information than just a sequence string or gene name might be required (e.g., reference to sequence information in some external database), therefore **Biosequence** is defined as a separate class.

A **Biosequence** object may represent actual complete sequence of the related **Elements**, actual partial sequence (several partial sequences can characterize a single **Element**), actual oligo sequence (several oligos can characterize a single **Element**), or a reference sequence (a sequence which is thought to represent the actual sequence(s) in the array element, e.g., if actual sequences are not known). A **Biosequence** object can also store clone or gene information if the related **Element** is described in terms of clones and/or genes, e.g., a pointer to the clone in a clone library database (if the probe has been obtained from a clone). If the sequence description is present in a biosequence database such as GenBank or EMBL bank, then it can be described by its accession number and an unambiguous database name.

For ESTs the links to the known EST clusters can be given. These cross-references will also provide the links into other databases, e.g., metabolic and functional databases that describe the gene.

A **Biosequence** can be related to several array **Elements**. The following information can be recorded using **Biosequence** class attributes:

- Type of the **Biosequence** object, e.g., complete sequence, partial sequence, oligo sequence, reference sequence, clone, gene.
- Actual sequence string, particularly if the sequence is not referenced within some database (through **description** → **Properties** → **ExternalReference**).
- Approximate length of the **Biosequence** if sequence is not known precisely.

5.4
Extract Preparation

5.4.1
SampleSource

Objects of this class represent organisms, cell lines, tissues or other sample sources used in the hybridization. Descriptions of **SampleSources** should include organism (NCBI taxonomy) and cell source and type [if derived from primary source(s)]. An additional property list may include: sex, age, development stage, organism part (tissue), animal/plant strain or line (if applicable), genetic variation (e.g., gene knock-out, transgenic variation), individual (if applicable), individual genetic characteristics (e.g., disease alleles, polymorphisms), disease state (or indication "normal"). A standard controlled vocabulary should ideally be used to characterize the **SampleSource**, but since there is no generally accepted nomenclature covering all cases, this may be only partly achievable.

One or more **SampleSource** objects can serve as sources for some **Treatment**, resulting either in another **SampleSource** (or **SampleSources**), or **Sample**(s), or combination of both.

5.4.2
Sample

This object describes the sample properties that can change during the experiment. One or more **Sample** objects can serve as sources for some **Treatment**, resulting in another **Sample** (or **Samples**). In addition, **Samples Treatments** can also have **SampleSources** as treatment sources.

Although initial **ExpressionValues** (see below) are calculated for each particular **Image**, the information we are really interested in is how each gene is expressed in each **Sample**. In ArrayExpressB we can trace **ExpressionValue** – **Sample** relations through the chain of classes **ExpressionValue** → **Expression-ValueSet** → **ImageAnalysis** → **Image** → **Hybridization** → **LabeledExtract** → **Labeling** → **Extract** → **Extraction** → **Sample**.

Whether a property should be recorded in the **Sample** or **SampleSource** class depends on the particular experiment. For example, in an experiment where gene expression changes during a progression of a cancer are studied, the description of the disease stage naturally belongs to **Sample** class (since it changes). In a drug-response type of study, where a particular same developmental stage is used, it belongs to **SampleSource**.

5.4.3
Treatment

Objects of this class can be used to describe in vivo treatments (then one or more **SampleSource** objects are sources for this treatment and one or more **Sample-Source** objects can be results of this treatment) or in vitro treatments, when source and/or result can be **Sample** object(s). Examples of such **Treatments** are "perform heat shock" (e.g., for yeast), "add compound", etc. A **Treatment** might be simply a delay, i.e., "wait 1 hour".

The **Treatment** attributes include: treatment type (e.g., heat shock, add compound), treatment compound, sample separation technique, **Treatment** protocol (free-text description).

5.4.4
Extraction and Extract

Objects of this class represent procedures for the isolation of the target molecules from a sample for their subsequent labeling and hybridization on an array. An **Extraction** uses one or more **Samples** and results in a single **Extract**. Description of how the **Extract** was produced from the **Sample** and prepared for **Hybridization** may include: cell rupture method, chemical extraction procedure (e.g., phenol, guanidine isothiocyanate), physical extraction procedure (e.g., centrifugation, poly-A beads, columns), whether total RNA, mRNA, or genomic DNA was extracted, amount of nucleic acids extracted, target amplification (RNA polymerases, PCR).

5.4.5
Labeling and LabeledExtract

Objects of the **Labeling** class represent instances of processes used to label **Extracts**, obtaining **LabeledExtracts**. Each **Labeling** uses exactly one **Extract** and produces exactly one **LabeledExtract**.

Description of labeling protocol can include: label used (e. g., Cy3, Cy5, 33P), amount of nucleic acids labeled, the labeling ratio (efficiency).

A **LabeledExtract** object is the substance that is hybridized to an array. Information on the process by which the **LabeledExtract** has been obtained can be traced back using relations **LabeledExtract** → **Labeling** → **Extract** → **Extraction** → **Sample** → **Treatment** → **SampleSource**.

5.5
Array Scanning and Analysis

5.5.1
Image

The next experimental steps after the hybridization typically are scanning of the hybridized array, which results in a digitized image (represented by **Image** class), followed by quantitation of the image (represented by **ImageAnalysis** class described below). The digitized images typically are not stored in the database, but they may be kept in a separate image archive (indexed in the database). If the image is not stored in the database, then the object of this class is a placeholder for annotation about the image and how it was produced. Some of **Image** attributes are: image URL – reference to the physical location of the image; protocol – used to record details of the scanning process (hardware, software, parameters, procedure); description that can include parsed header of the TIFF file, including laser power, spatial resolution, pixel space, PMT voltage.

5.5.2
ImageAnalysis

Objects of this class correspond to instances of scanned microarray image analysis. A single **Image** can be analyzed several times using different protocols, and several **Images** (from the same **Hybridization**) can be used for single analysis. Result of an **ImageAnalysis** is an **ExpressionValueSet**.

For description of the analysis characteristics, **protocol** attribute should be used. The software and parameters used are specified, as well the quality assessment of the overall image (if available) and any normalization that has been applied before the final output.

5.5.3
ExpressionValueSet, ExpressionValue and ExpressionValueType

A single **ExpressionValueSet** can be obtained from **ImageAnalysis**. An **ExpressionValueSet** contains many **ExpressionValues**. Each **ExpressionValue** may be linked to one or more **Elements**. There is an indirect link to **Sample** through **ExpressionValueSet** → **ImageAnalysis** → **Image** → **Hybridization** → **LabeledExtract** → **Labeling** → **Extract** → **Extraction** → **Sample**.

Although the numerical measurement data depends on the particular image analysis algorithm, usually we can distinguish two types of **ExpressionValues**. The first type includes raw data that is a direct output by the image analysis software (for each spot, mean intensity, mean background, median intensity, median background, number of pixels in the spot, standard deviation of pixel intensities, etc., depending on the particular software). The second type is consolidated values that can be directly used to characterize the expression level of a particular gene (or reliability of a particular expression measurement), such as background-subtracted mean intensity for a spot. The raw information can be useful in assessing the reliability of the final expression level measurements. For example, the correlation between the intensities of the pixels in a spot is a possible reliability indicator. Ideally, we would like to be able to deduce the measurement error level from these numbers; unfortunately, currently, there are no standard reliability measures accepted (this would require an agreement on the definition of the standard error based on the analysis of only one spot). We envisage that, in the future, methods and standards for the measurement of quality assessment will be developed. In this case it may be sufficient to store only the measurement value and the quality assessment for each gene.

The attributes of the **ExpressionValue** class are:

- actual numerical value; units and other relevant information are given in the related **ExpressionValueType** object;
- quality score of the **ExpressionValue** (if applicable for the particular **ExpressionValueType**).

Categories of **ExpressionValues** are objects of expression value type class. Attributes include:

- name of the **ExpressionValueType** object (e.g., "channel1 intensity", "background pixel count"),
- measurement units,
- type of the **qualityScore** attribute of associated **ExpressionValue** objects (if applicable),
- measurement units for **qualityScores**.

5.6
Common Constructs

Pieces of information, such as protocols and external references, are common for all areas of the expression data domain. For example, protocols are used to describe **ImageAnalysis**, as well as **Hybridizations**. Although these are rather different kinds of protocols in terms of actions they describe, from the viewpoint of the information structure they are rather similar, therefore a common **Protocol** class is introduced.

5.6.1
Description

This is a catchall class to store annotation recorded in a free-text description as well as in a controlled manner using name-value pair lists for **Properties**.

5.6.2
Protocol

Objects of this class are referenced from **protocol** attribute of classes **Treatment, Extraction, Labeling, Hybridization, Image, ImageAnalysis, ArrayType, ExpressionValueType. Protocol** is a subclass of **Description**, i. e., it can consist of free-text description and properties. In the future more structured description of **Protocols** can be envisaged (involving, e. g., protocol steps).

5.6.3
Property

The name-value pair, e. g., age = 29 years; sex = male; development_stage = adult. Value is stored in attribute **value**, while property name can be found in the related **PropertyType** object, where, if necessary, a **unit** for the value may also be specified. **Properties** are grouped to form **Descriptions**. A **Property** can have an **ExternalReference** attached, either documenting **Property** value or acting as an external **Property** value (then attribute **value** should be empty).

5.6.4
PropertyType

Objects of this class determine what types of properties are available in descriptions of objects of various ArrayExpress classes. Attributes of this class include: property name, measurement units, free-text description, list of class names determining which objects can have properties of this type.

5.6.5
ExternalReference

Any object that is an **ExternalReference** (i. e., **LiteratureReference, DatabaseReference, OntologyReference** and **WebReference**) can be associated with a **Property** that is a part of some **Description**. Such **ExternalReference** either documents how the **value** attribute of the **Property** object was obtained, or essentially provides the value (then **Property** → **value** should be empty).

5.6.6
LiteratureReference

These can be references to publications that can be attached to any **Description** (since **LiteratureReference** is a subclass of **ExternalReference**). The only attribute is **pubMedID**; for **LiteratureReferences** that are not in PubMed, **description** attribute (of the superclass **ExternalReference**) should be used. Typically, **LiteratureReferences** will be used in **Descriptions** of **Experiments** and **ArrayTypes**.

5.6.7
DatabaseReference

This class is a subclass of **ExternalReference** class. An object of the **DatabaseReference** class represents either some external database as a whole, or a particular entry in that database. The most typical use of **DatabaseReferences** within ArrayExpress will be linking **Biosequence** objects to corresponding database entries in external databases. A **DatabaseReference** object contains a unique identifier of the database, database URL, and an identifier that permits unique identification of the relevant object within the database.

5.6.8
OntologyReference

In order to describe ArrayExpress objects (especially **SampleSources** and **Samples**), **Description** type objects will be used, incorporating both free text and controlled text through property name – property value pairs. It is advisable to use a reference to an external ontology where appropriate. Objects of the class **OntologyReference** link together property values with external ontologies.

An **OntologyReference** object contains ontology URL (it is assumed that an external ontology has to be encoded in some XML-compliant format and should have a URL) and a reference to a specific element of the ontology.

5.6.9
WebReference

An object of this class can be used as a general way to reference some Web resource (which is not a database, ontology or article).

6
Complete Model – ArrayExpressC

There are experimental setups for which the Basic model is not sufficiently detailed. For instance, ArrayExpressB does not allow the capture of **ExpressionValue** when it is related to more than one array **Element**. Also there is no way to capture relations between several **ExpressionValueSets**. For instance, an **ExpressionValueSet** can be obtained by averaging several **ExpressionValueSets** each obtained from different replicate **Hybridizations**. ArrayExpressC provides means to deal with these cases in a technology and expression data analysis independent way. There are also several other differences between ArrayExpressB and Array-ExpressC that are described below.

6.1
Experiment as a Whole

An **Experiment** can be related to one or more other **Experiments** through the recursive association **relates**. A **ControlElement** in this model is used to capture the situation when there is an **Element** on the **Array** that for a particular **Hybridization** is expected to yield some specific expression value (of type **ExpressionValueType**).

6.2
Hybridization

The **Array** class is used to represent a particular instance of an array used for hybridization, as opposed to **ArrayType**. This allows the description of an experiment where several **Hybridizations** can be performed on the same **Array** (i.e., instance of **ArrayType**). Recursive **replicate** relation can relate several **Hybridizations** regarded as replicates of the same conditions. **Hybridization** class is a subclass of a more general **Event** class (described below).

6.3
Describing Arrays and Array Types

6.3.1
Array and ArrayBatch

An **Array** belongs to an **ArrayBatch**. All **Arrays** derived from a certain **ArrayType** are physically identical, but each has a unique array ID (e.g., a barcode as used in the laboratory for tracking purposes) that differentiates it from the others.

In the ArrayExpressB model, **Hybridization** is directly associated with **ArrayType**, assuming that it is not important whether several **Hybridizations** are performed on the same instance of some **ArrayType** or on several instances.

Arrays of the same **ArrayType** can perform differently depending on which **ArrayBatch** they belong to. An **ArrayBatch** object represents a set of **Arrays** belonging to the same **ArrayType** printed/synthesized using the same raw materials.

6.3.2
Element

Each **Element** is associated with exactly one **Reporter**. **Element** class is a subclass of ExpressionValueDimension class. If an **ExpressionValue** object is related to exactly one **Element,** then it will be associated with it through the **ExpressionValue** → ExpressionValueDimension association.

Elements on arrays are sometimes grouped within ElementBlocks, contiguous groups of Elements in terms of array layout, that in turn can be grouped to form larger ElementBlocks. ElementGroup objects represent groups of Elements that share the same type and generation protocol. Elements within a group can differ only in position and Reporters that the Elements represent. ElementGroup attributes include: element type, e. g., synthesized oligo-nucleotides, PCR products, plasmids, colonies, whether single or double stranded, expected dimensions of all elements within the group, according to Element generation protocol, reference to Protocol used for synthesizing Elements within the group, attachment type, e. g., covalent, ionic, hydrophobic, other.

6.3.3
Reporter

Reporter class objects represent sequences that are attached to arrays as Elements. Each Reporter object can have several related Element objects (either on the same ArrayType or on several ArrayTypes). Each Reporter is described by one or more (e. g., in case of several known partial sequences) Biosequence objects.

Reporter class is introduced to explicitly model distinction between physical Elements on arrays and "logical" elements that are characterized by one or more Biosequence class objects. Note that there is no standard terminology for these notions. What we call Reporters are sometimes called probes or targets, in which case LabeledExtracts in our model are called targets or probes, respectively. To avoid possible confusion resulting from the opposite use of terms "probe" and "target" we avoid them entirely.

6.4
Extract Preparation

6.4.1
Biomaterial

Objects of this class can represent materials used in microarray experiments in various stages of extraction: source, sample, extract, extract after labeling. Biomaterial is a superclass of SampleSource, Sample, Extract and LabeledExtract classes.

Biomaterials can be transformed by Events, resulting in new Biomaterials (or new states of the same biomaterials that should be represented by new objects of Biomaterial class). Two associations link Biomaterial and Event classes, one used for linking Biomaterial(s) as source(s) for some Event, another for linking Biomaterial(s) as result(s) of some Event.

There are associations connecting subclasses of Biomaterial to subclasses of Event (Treatment, Extraction, Labeling, Hybridization), forming a typical chain of actions performed during microarray experiments. Biomaterial and Event class objects as more abstract, together with abstract Biomaterial → Event associations, should be used when this typical chain is not sufficient.

6.4.2
Event

Labeled extracts used in microarray experiments can be traced starting from extract source (organism), through sample to extract and labeled extract. The single superclass for all these objects is **Biomaterial**. Objects of the **Biomaterial** class can be thought of as nodes in an acyclic directed graph, with arcs being **Events** that transform sets of **Biomaterials** into other sets of **Biomaterials**. In the ArrayExpressC model, a **Hybridization** is a subclass of the **Event** class.

Event objects can be composed of several other **Event** objects through **event-Composition** recursive aggregation. Therefore, complex treatments can be described in terms of simpler treatments, and it is not necessary for each simple **Event** object to introduce a new **Biomaterial** class object representing a biomaterial state after that simple **Event** and before other successive **Events**.

Each **Event** has its name, which ideally should characterize the treatment (e. g., heat shock, cold shock, food deprivation, add chemical, dry, centrifuge, resuspend, wait). Since **Event** class objects can document many types of biological manipulations, **protocol** for an **Event** can include various types of descriptions (see also descriptions of **Event** subclasses):

- How the source was perturbed and treated in order to produce the sample (e. g., apply heat shock). It is possible to subcategorize various treatments into the following groups: in vivo treatments (organism or individual treatments), in vitro treatments (cell culture conditions), and separation technique (e. g., none, trimming, microdissection, FACS).
- How an extract was produced from a sample and prepared for hybridization, e. g., extraction method, whether total RNA, mRNA, or genomic DNA was extracted, exogenous sequences (spikes) added (for each spike added: type, name/clone id, quantity added), target amplification (RNA polymerases, PCR).
- How the labeling was performed.
- How the labeled extract was hybridized to the array.

6.4.3
Sample

In ArrayExpressC, **Sample** class is a subclass of **ExpressionValueDimension** class, thus **Sample** objects can be related to **ExpressionValue** objects.

6.5
Array Scanning and Analysis

In the ArrayExpressC model, the description of the array scanning and analysis allows for many-to-many relationships between the array elements and expression values.

6.5.1
ImageAnalysis

The many-to-one association between **ImageAnalysis** and **Hybridization** emphasizes that several **Images** can be used for single **ImageAnalysis** only if they are obtained from the same **Hybridization**. **ImageAnalysis** is a subclass of **ExpressionValueDimension**; therefore, **ExpressionValues** can be directly related to **ImageAnalysis** objects.

6.5.2
ExpressionValueSet

In ArrayExpressB an association between **ImageAnalysis** and **ExpressionValueSet** was introduced. However, an **ExpressionValueSet** can also be a result of some **Transformation** (described below). **ExpressionValueSet** contains **ExpressionValues** that have links with several **ExpressionValueDimension** objects. There is a direct association between **ExpressionValueSet** and **ExpressionValueDimension**. A **Transformation** can consume one or more **ExpressionValueSets** and produce one **ExpressionValueSet**. Each **ExpressionValueSet** is a product of no more than one **Transformation**, but it can serve as a source for several **Transformations**.

6.5.3
ExpressionValue

In ArrayExpressC an **ExpressionValue** object can have links to several **ExpressionValueDimension** objects. **ExpressionValueDimension** is an abstraction of all classes that can be directly or indirectly linked to **ExpressionValues**.

6.5.4
ExpressionValueDimension

ExpressionValueDimension is an abstraction of all classes that can be directly or indirectly linked to **ExpressionValues**. Subclasses of **ExpressionValueDimension** are **ExpressionValueType, Element, CompositeElement, Sample, CompositeSample, ImageAnalysis**. Such abstraction was introduced to permit easier modeling of **Transformations** (described below).

These dimensions are not necessarily independent; since, e. g., the related **Sample** object depends on the related **ImageAnalysis** object, some **ExpressionValueTypes** are meaningful only for single **Samples**, some only for **CompositeSample** (e. g., ratio), etc.

Each **ExpressionValue** can have up to four related **ExpressionValueDimension** objects:

- **ExpressionValueType** (mandatory).
- **Element** or **CompositeElement**, i. e., which **Element**(s) the **ExpressionValue** is related to.
- **ImageAnalysis**, allowing to track which **Image**(s) served as a source for this **ExpressionValue** and what protocol was used. This is an optional dimension,

since an **ExpressionValue** might be obtained by some **Transformation** involving several primary **ExpressionValueSets** resulting from several **Image-Analysis** objects.

– **Sample** or **CompositeSample**. This dimension depends on **ImageAnalysis** dimension, but was added in the model to stress that, although technically **ExpressionValues** are obtained through **ImageAnalysis**, meaningful information for further analysis is which **Samples** these **ExpressionValues** characterize.

There are two relationships between **ExpressionValueDimension** class and **Transformation** class. **Transformations** transform **ExpressionValueSets** yielding new **ExpressionValueSets**, and they can introduce new values for expression value dimensions (i.e., new **ExpressionValueDimension** objects). For example, each instance of the ratio transformation transforms two raw image **ExpressionValueSets** into a ratio **ExpressionValueSet**, introducing some new **ExpressionValueDimension** objects: **ExpressionValueType** "ratio" and a new **CompositeSample** object. For a **Transformation** there may be one or more **sourceDimensions**, and a **Transformation** may yield one or more **resultDimensions**.

6.5.5
ExpressionValueType

For an **ExpressionValueType** object, a **description** attribute can be used to describe how this **ExpressionValueType** relates to its "primary" **ExpressionValueTypes**, if it is obtained by some **Transformation**.

6.5.6
CompositeElement

Some of the **ExpressionValues** can be computed using primary data (from image scan) originating from more than one **Element** (e.g., if there are several **Elements** on an array representing the same gene, or if the same **Reporter** is spotted as many **Elements** on a single array). Then one of the **ExpressionValueDimension** objects associated with this **ExpressionValue** will actually be an object of **CompositeElement** class (which is a subclass of **ExpressionValueDimension**), and this **CompositeElement** will be associated with all relevant **Elements**.

6.5.7
CompositeSample

Some of the **ExpressionValues** are related to more than one **Sample** (the most typical example being usual ratios). Then one of the **ExpressionValueDimension** objects associated with such an **ExpressionValue** will actually be an object of **CompositeSample** class (which is a subclass of **ExpressionValueDimension**), and this **CompositeSample** will be associated with all relevant **Samples**.

6.5.8
Transformation

Used for tracking processing of **ExpressionValueSets**. Each **Transformation** consumes one or more **ExpressionValueSets** and produces an **ExpressionValueSet**. For example, a **Transformation** can average several **ExpressionValueSets** obtained from replicate **Hybridizations,** or normalize an **ExpressionValueSet**. A **Transformation** can be a composition of several simpler **Transformations**.

A **Transformation** can also create new **ExpressionValueDimensions**. If we regard an **ExpressionValueSet** as a (for simplicity two-dimensional) matrix, then a **Transformation** can create new columns and rows in this matrix (e.g., a ratio column or an average column). **SourceDimensions** for this **Transformation** will be all **ExpressionValueDimensions** (i.e., rows or columns) containing **ExpressionValues** used to compute **ExpressionValues** of the new dimension(s).

6.6
Common Constructs

An additional association between **Description** and **Property** classes (called **references**) is added to allow **Property** objects belonging to one **Description** object to reference other **Description** objects. Such reference might be necessary, e.g., when there are several **Descriptions** of scanning hardware and we want to reference from some scanning software **Description** which hardware is compatible with this software. A **Property** belongs to one **Description,** but it can reference other **Descriptions**. Thus, **Description** and **Property** class objects can encode complex structures, if needed.

There is an association between **Property** and **Submitter** classes. It can be used when database objects are annotated by database users other than (object) submitters; then **source** association points to the **Submitter** of the **Property**.

Objects of the **StandardProtocol** class are referenced by **Protocol** class objects. A **StandardProtocol** object can be thought of as a protocol type, while a **Protocol** object is a protocol application instance. **StandardProtocol** is just a subclass of **Description**; however, in the future a more structured description of **StandardProtocols** can be envisaged (involving, e.g., protocol steps).

7
Implementation Issues

7.1
Database Implementation

The ArrayExpressB and ArrayExpressC object models presented here should not be regarded as optimal for relational database implementation using straightforward object model to entity-relationship model mapping. For example, creating the ExpressionValue table with a single expression value in each row may not lead to efficient implementation. Instead, single values should be grouped into larger chunks and stored in the database this way. Object model should be made stable, whereas underlying relational database implementation can change more often. Rational rose MDL files are available from (http://www.ebi.ac.uk/array-express/).

ArrayExpress implementation at EBI will run on an Oracle 8i platform; however, database schema will be easily portable to other RDBMSs. The supported data import format will be MAML, a MIAME-compliant XML language[1]. Images will not be stored inside the database, they will be archived on tapes or direct access media such as CD-R or DVD-R.

7.2
Data Queries, Mining and Visualization

The ultimate purpose of the database is not just to archive the data, but to make it accessible for researchers in convenient forms. This includes the necessity to implement queries combining various properties of the data and annotations. For instance, we might look for genes in the database with expression profiles similar to a given one.

An object model-based query interface to ArrayExpress is being developed. The Web interfaces for predefined types of queries will be provided on top of the general query mechanism. For the database to be used for efficient data mining and interactive visualization, extensive optimization may be required, e. g., by tuning of table indexing or producing a denormalized database schema for some parts of the database.

8
Discussion

One essential question is whether to store the raw image data. Currently, there are rather opposite views in the microarray community, ranging from the view that without storing the images the database would be unreliable, to the view that storing the images in any form is cost-ineffective.

When designing a database for storing information originating from physical experiments, we have to decide on the desired level of processing of the raw data before they are stored. For example, in DNA sequencing, we could treat the fluorescent traces as the raw data and the sequences as the processed data. However, a sequence is a discrete object and the error rate in determining the actual sequence is low. Therefore, the advantage of storing the sequences instead of traces is obvious (although lately the need to access the trace information as well is becoming increasingly evident). In contrast to sequence data, gene expression level measurements do not represent discrete objects, but are approximate numbers (unless the precision of the measurements was such that we could determine the precise mRNA molecule count). Moreover, the noise level in gene expression studies is much higher than in sequencing. In this case, reducing the images to numeric values may result in loss of valuable information. On the other hand, this

---

[1] See (http://sourceforge.net/projects/mged/).

numerical representation of images is necessary for higher-level data analysis, e. g., for clustering genes according to their expression profiles. The pros and cons for storing the images or not, or intermediate options, ranging from archiving the images separately from the database (for instance, on CD-ROMs), to storing highly compressed images suited only for "eye-balling" (which do not allow their re-analysis), should be weighed up by the individual laboratories and the most appropriate decision made.

To transform the raw experimental data into a gene expression matrix, many nontrivial steps are needed, of which the image analysis is only one. For example, a gene may be spotted several times on an array, using either the same or different parts of the gene sequence. The expression level measurements of different spots representing the same gene in the same hybridization experiment may be different, and sometimes even contradictory. To obtain the conceptual table of gene expression values for different conditions, these conflicts should be resolved and the information consolidated. Even if a good consolidation procedure can be found, some experimental information is lost, unless all the expression measurements regarding the same gene are kept in the database. It may be even more difficult to consolidate information from different experiments, where different names or accession numbers are used for the same gene. The last problem is not specific for a gene expression database, but is haunting bioinformatics in general. It does not seem practical to tie the solution of this general problem with the implementation of the gene expression database.

Another question that has to be considered by database developers is where to make the compromise between the database complexity and flexibility. The reason why the presented ArrayExpress model is so complicated is the need to capture data from different microarray platforms in a single model. Laboratories using only one technology platform may simplify the model. On the other hand, the microarray technology is still rapidly developing, and a more generic model may have the advantage that it may not have to be changed when the technology is upgraded.

The data model is able to capture detailed experiment annotation. Annotating the experiments in such detail may be regarded as an unnecessary burden by some laboratories. In the current gene expression databases, some of the samples are annotated rather briefly, for instance as 'control'. Laboratories should consider the tradeoffs of investing resources in annotation of their experiments in more or less detail, and being able to use the obtained data at a later date.

Acknowledgements. The authors wish to thank the MGED group for its contribution in developing the requirements for the database. In particular we would like to thank Paul Spellman, from University of California at Berkeley, Alex Lash at NCBI, Harry Mangalam and Jason Stewart at NCGR, Michael Bittner at NHGRI, and Frank Holstege at the University of Utrecht. We would also like to thank the Database Applications group at the EBI, and Nicole Redaschi in particular. The development of the ArrayExpress prototype is supported by Incyte. The ArrayExpress object model was influenced by Incyte's model that was developed as a part of Incyte's Genomic Knowledge Platform initiative, aimed towards unifying genomics-related information sources and tools by providing a single comprehensive object model and implementing necessary middleware components. The authors wish to thank Lee Grover, Frank Russo and Greg Pelz for their contribution.

9
References

1. The Chipping Forecast (1999) *Nature Genetics* 21 (Special Supplement)
2. Fowler M, Scott K, Booch G (1999) UML distilled, 2nd edn. A brief guide to the standard object modeling language, The Addison-Wesley Object Technology Series

Received: June 2001

References

Author Index Volume 51–77

Author Index Volumes 1–50 see Volume 50

Ackermann, J.-U. see Babel, W.: Vol. 71, p. 125

Adam, W., Lazarus, M., Saha-Möller, C. R., Weichhold, O., Hoch, U., Häring, D., Schreier, Ü.: Biotransformations with Peroxidases. Vol. 63, p. 73

Akhtar, M., Blanchette, R. A., Kirk, T. K.: Fungal Delignification and Biochemical Pulping of Wood. Vol. 57, p. 159

Allan, J. V., Roberts, S. M., Williamson, N. M.: Polyamino Acids as Man-Made Catalysts. Vol. 63, p. 125

Allington, R. W. see Xie, S.: Vol. 76, p. 87

Al-Rubeai, M.: Apoptosis and Cell Culture Technology. Vol. 59, p. 225

Al-Rubeai, M. see Singh, R. P.: Vol. 62, p. 167

Alsberg, B. K. see Shaw, A. D.: Vol. 66, p. 83

Antranikian, G. see Ladenstein, R.: Vol. 61, p. 37

Antranikian, G. see Müller, R.: Vol. 61, p. 155

Archelas, A. see Orru, R. V. A.: Vol. 63, p. 145

Argyropoulos, D. S.: Lignin. Vol. 57, p. 127

Arnold, F. H., Moore, J. C.: Optimizing Industrial Enzymes by Directed Evolution. Vol. 58, p. 1

Autuori, F., Farrace, M. G., Oliverio, S., Piredda, L., Piacentini, G.: "Tissie" Transglutaminase and Apoptosis. Vol. 62, p. 129

Azerad, R.: Microbial Models for Drug Metabolism. Vol. 63, p. 169

Babel, W., Ackermann, J.-U., Breuer, U.: Physiology, Regulation and Limits of the Synthesis of Poly(3HB). Vol. 71, p. 125

Bajpai, P., Bajpai, P. K.: Realities and Trends in Emzymatic Prebleaching of Kraft Pulp. Vol. 56, p. 1

Bajpai, P., Bajpai, P. K.: Reduction of Organochlorine Compounds in Bleach Plant Effluents. Vol. 57, p. 213

Bajpai, P. K. see Bajpai, P.: Vol. 56, p. 1

Bajpai, P. K. see Bajpai, P.: Vol. 57, p. 213

Barut, M. see Strancar, A.: Vol. 76, p. 49

Bárzana, E.: Gas Phase Biosensors. Vol. 53, p. 1

Bazin, M. J. see Markov, S. A.: Vol. 52, p. 59

Bellgardt, K.-H.: Process Models for Production of β-Lactam Antibiotics. Vol. 60, p. 153

Beppu, T.: Development of Applied Microbiology to Modern Biotechnology in Japan. Vol. 69, p. 41

Berovic, M. see Mitchell, D. A.: Vol. 68, p. 61

Beyeler, W., DaPra, E., Schneider, K.: Automation of Industrial Bioprocesses. Vol. 70, p. 139

Beyer, M. see Seidel, G.: Vol. 66, p. 115

Bhatia, P. K., Mukhopadhyay, A.: Protein Glycosylation: Implications for in vivo Functions and Thereapeutic Applications. Vol. 64, p. 155

Bisaria, V. S. see Ghose, T. K.: Vol. 69, p. 87

Blanchette R. A. see Akhtar, M.: Vol. 57, p. 159

Bocker, H., Knorre, W.A.: Antibiotica Research in Jena from Penicillin and Nourseothricin to Interferon. Vol. 70, p. 35

de Bont, J.A.M. see van der Werf, M. J.: Vol. 55, p. 147

van den Boom, D. see Jurinke, C.: Vol. 77, p. 57

Brainard, A. P. see Ho, N. W. Y.: Vol. 65, p. 163

Brazma, A., Sarkans, U., Robinson, A., Vilo, J., Vingron, M., Hoheisel, J., Fellenberg, K.: Microarray Data Representation, Annotation and Storage. Vol. 77, p. 113

Breuer, U. see Babel, W.: Vol. 71, p. 125

Broadhurst, D. see Shaw, A. D.: Vol. 66, p. 83

Bruckheimer, E. M., Cho, S. H., Sarkiss, M., Herrmann, J., McDonell, T. J.: The Bcl-2 Gene Family and Apoptosis. Vol 62, p. 75

Brüggemann, O.: Molecularly Imprinted Materials – Receptors More Durable than Nature Can Provide. Vol. 76, p. 127

Buchert, J. see Suurnäkki, A.: Vol. 57, p. 261

Bungay, H. R. see Mühlemann, H. M.: Vol. 65, p. 193

Bungay, H.R., Isermann, H. P.: Computer Applications in Bioprocessin. Vol. 70, p. 109

Büssow, K. see Eickhoff, H.: Vol. 77, p. 103

Byun, S. Y. see Choi, J. W.: Vol. 72, p. 63

Cantor, C.R. see Jurinke, C.: Vol. 77, p. 57

Cao, N. J. see Gong, C. S.: Vol. 65, p. 207

Cao, N. J. see Tsao, G. T.: Vol. 65, p. 243

Carnell, A. J.: Stereoinversions Using Microbial Redox-Reactions. Vol. 63, p. 57

Cen, P., Xia, L.: Production of Cellulase by Solid-State Fermentation. Vol. 65, p. 69

Chang, H. N. see Lee, S. Y.: Vol. 52, p. 27

Cheetham, P. S. J.: Combining the Technical Push and the Business Pull for Natural Flavours.Vol. 55, p. 1

Chen, Z. see Ho, N. W. Y.: Vol. 65, p. 163

Cho, S. H. see Bruckheimer, E. M.: Vol. 62, p. 75

Cho, G.H. see Choi, J. W.: Vol 72, p. 63

Choi, J. see Lee, S. Y.: Vol. 71, p. 183

Choi, J.W., Cho, G.H., Byun, S. Y., Kim, D.-I.: Integrated Bioprocessing for Plant Cultures. Vol. 72, p. 63

Christensen, B., Nielsen, J.: Metabolic Network Analysis – A Powerful Tool in Metabolic Engineering. Vol. 66, p. 209

Christians, F. C. see McGall, G.H.: Vol. 77, p. 21

Chui, G. see Drmanac, R.: Vol. 77, p. 75

Ciaramella, M. see van der Oost, J.: Vol. 61, p. 87

Contreras, B. see Sablon, E.: Vol. 68, p. 21

Cordero Otero, R. R. see Hahn-Hägerdal, B.: Vol. 73, p. 53

Cornet, J.-F., Dussap, C. G., Gros, J.-B.: Kinetics and Energetics of Photosynthetic Micro-Organisms in Photobioreactors. Vol. 59, p. 153

da Costa, M. S., Santos, H., Galinski, E. A.: An Overview of the Role and Diversity of Compatible Solutes in Bacteria and Archaea. Vol. 61, p. 117

Cotter, T. G. see McKenna, S. L.: Vol. 62, p. 1

Croteau, R. see McCaskill, D.: Vol. 55, p. 107

Danielsson, B. see Xie, B.: Vol. 64, p. 1

DaPra, E. see Beyeler, W.: Vol. 70, p. 139

Darzynkiewicz, Z., Traganos, F.: Measurement of Apoptosis. Vol. 62, p. 33

Davey, H. M. see Shaw, A. D.: Vol. 66, p. 83

Dean, J. F. D., LaFayette, P. R., Eriksson, K.-E. L., Merkle, S. A.: Forest Tree Biotechnolgy. Vol. 57, p. 1

Demain, A. L., Fang, A.: The Natural Functions of Secondary Metabolites. Vol. 69, p. 1

Diaz, R. see Drmanac, R.: Vol. 77, p. 75

Dochain, D., Perrier, M.: Dynamical Modelling, Analysis, Monitoring and Control Design for Nonlinear Bioprocesses. Vol. 56, p. 147

Drmanac, R., Drmanac, S., Chui, G., Diaz, R., Hou, A., Jin, H., Jin, P., Kwon, S., Lacy, S., Moeur, B., Shafto, J., Swanson, D., Ukrainczyk, T., Xu, C., Little, D.: Sequencing by Hybridization (SBH): Advantages, Achievements, and Opportunities. Vol. 77, p. 75

Drmanac, S. see Drmanac, R.: Vol. 77, p. 75

Du, J. see Gong, C. S: Vol. 65, p. 207

Du, J. see Tsao, G. T.: Vol. 65, p. 243

Dueser, M. see Raghavarao, K. S. M. S.: Vol. 68, p. 139

Dussap, C. G. see Cornet J.-F.: Vol. 59, p. 153

Dutta, N. N. see Ghosh, A. C.: Vol. 56, p. 111

Dutta, N. N. see Sahoo, G. C.: Vol. 75, p. 209

Dynesen, J. see McIntyre, M.: Vol. 73, p. 103

Eggeling, L., Sahm, H., de Graaf, A. A.: Quantifying and Directing Metabolite Flux: Application to Amino Acid Overproduction. Vol. 54, p. 1

Eggeling, L. see de Graaf, A. A.: Vol. 73, p. 9

Eggink, G., see Kessler, B.: Vol. 71, p. 159

Eggink, G., see van der Walle, G. J. M.: Vol. 71, p. 263

Ehrlich, H. L. see Rusin, P.: Vol. 52, p. 1

Eickhoff, H., Konthur, Z., Lueking, A., Lehrach, H., Walter, G., Nordhoff, E., Nyarsik, L., Büssow, K.: Protein Array Technology: The Tool to Bridge Genomics and Proteomics. Vol. 77, p. 103

Elias, C. B., Joshi, J. B.: Role of Hydrodynamic Shear on Activity and Structure of Proteins. Vol. 59, p. 47

Elling, L.: Glycobiotechnology: Enzymes for the Synthesis of Nucleotide Sugars. Vol. 58, p. 89

Eriksson, K.-E. L. see Kuhad, R. C.: Vol. 57, p. 45

Eriksson, K.-E. L. see Dean, J. F. D.: Vol. 57, p. 1

Faber, K. see Orru, R. V. A.: Vol. 63, p. 145

Fang, A. see Demain, A. L.: Vol. 69, p. 1

Farrace, M. G. see Autuori, F.: Vol. 62, p. 129

Farrell, R. L., Hata, K., Wall, M. B.: Solving Pitch Problems in Pulp and Paper Processes. Vol. 57, p. 197

Fellenberg, K. see Brazma, A.: Vol. 77, p. 113

Fiechter, A.: Biotechnology in Switzerland and a Glance at Germany. Vol. 69, p. 175

Fiechter, A. see Ochsner, U. A.: Vol. 53, p. 89

Foody, B. see Tolan, J. S.: Vol. 65, p. 41

Fréchet, J. M. J. see Xie, S.: Vol. 76, p. 87

Freitag, R., Hórvath, C.: Chromatography in the Downstream Processing of Biotechnological Products. Vol. 53, p. 17

Furstoss, R. see Orru, R. V. A.: Vol. 63, p. 145

Galinski, E. A. see da Costa, M. S.: Vol. 61, p. 117

Gàrdonyi, M. see Hahn-Hägerdal, B.: Vol. 73, p. 53

Gatfield, I. L.: Biotechnological Production of Flavour-Active Lactones. Vol. 55, p. 221

Gemeiner, P. see Stefuca, V.: Vol. 64, p. 69

Gerlach, S. R. see Schügerl, K.: Vol. 60, p. 195

Ghose, T. K., Bisaria, V. S.: Development of Biotechnology in India. Vol. 69, p. 71

Ghosh, A. C., Mathur, R. K., Dutta, N. N.: Extraction and Purification of Cephalosporin Antibiotics. Vol. 56, p. 111

Ghosh, P. see Singh, A.: Vol. 51, p. 47

Gilbert, R. J. see Shaw, A. D.: Vol. 66, p. 83

Gill, R. T. see Stephanopoulos, G.: Vol. 73, p. 1

Gomes, J., Menawat, A. S.: Fed-Batch Bioproduction of Spectinomycin. Vol. 59, p. 1

Gong, C. S., Cao, N. J., Du, J., Tsao, G. T.: Ethanol Production from Renewable Resources. Vol. 65, p. 207

Gong, C. S. see Tsao, G. T.: Vol. 65, p. 243

Goodacre, R. see Shaw, A. D.: Vol. 66, p. 83

de Graaf, A. A., Eggeling, L., Sahm, H.: Metabolic Engineering for L-Lysine Production by *Corynebacterium glutamicum.* Vol. 73, p. 9

de Graaf, A.A. see Eggeling, L.: Vol. 54, p. 1

de Graaf, A.A. see Weuster-Botz, D.: Vol. 54, p. 75

de Graaf, A.A. see Wiechert, W.: Vol. 54, p. 109

Grabley, S., Thiericke, R.: Bioactive Agents from Natural Sources: Trends in Discovery and Application. Vol. 64, p. 101

Griengl, H. see Johnson, D. V.: Vol. 63, p. 31

Gros, J.-B. see Larroche, C.: Vol. 55, p. 179

Gros, J.-B. see Cornet, J. F.: Vol. 59, p. 153

Guenette M. see Tolan, J. S.: Vol. 57, p. 289

Gutman, A. L., Shapira, M.: Synthetic Applications of Enzymatic Reactions in Organic Solvents. Vol. 52, p. 87

Hahn-Hägerdal, B., Wahlbom, C. F., Gárdonyi, M., van Zyl, W. H., Cordero Otero, R. R., Jönsson, L. J.: Metabolic Engineering of *Saccharomyces cerevisiae* for Xylose Utilization. Vol. 73, p. 53

Haigh, J. R. see Linden, J. C.: Vol. 72, p. 27

Hall, D. O. see Markov, S. A.: Vol. 52, p. 59

Hall, P. see Mosier, N. S.: Vol. 65, p. 23

Hammar, F.: History of Modern Genetics in Germany. Vol. 75, p. 1

Hannenhalli, S., Hubbell, E., Lipshutz, R., Pevzner, P. A.: Combinatorial Algorithms for Design of DNA Arrays. Vol. 77, p. 1

Haralampidis, D., Trojanowska, M., Osbourn, A. E.: Biosynthesis of Triterpenoid Saponins in Plants. Vol. 75, p. 31

Häring, D. see Adam, E.: Vol. 63, p. 73

Harvey, N. L., Kumar, S.: The Role of Caspases in Apoptosis. Vol. 62, p. 107

Hasegawa, S., Shimizu, K.: Noninferior Periodic Operation of Bioreactor Systems. Vol. 51, p. 91

Hata, K. see Farrell, R. L.: Vol. 57, p. 197

van der Heijden, R. see Memelink, J.: Vol. 72, p. 103

Hein, S. see Steinbüchel, A.: Vol. 71, p. 81

Hembach, T. see Ochsner, U. A.: Vol. 53, p. 89

Henzler, H.-J.: Particle Stress in Bioreactor. Vol. 67, p. 35

Herrmann, J. see Bruckheimer, E. M.: Vol. 62, p. 75

Hill, D. C., Wrigley, S. K., Nisbet, L. J.: Novel Screen Methodologies for Identification of New Microbial Metabolites with Pharmacological Activity. Vol. 59, p. 73

Hiroto, M. see Inada, Y.: Vol. 52, p. 129

Ho, N. W. Y., Chen, Z., Brainard, A. P. Sedlak, M.: Successful Design and Development of Genetically Engineering Saccharomyces Yeasts for Effective Cofermentation of Glucose and Xylose from Cellulosic Biomass to Fuel Ethanol. Vol. 65, p. 163

Hoch, U. see Adam, W.: Vol. 63, p. 73

Hoheisel, J. see Brazma, A.: Vol. 77, p. 113

Holló, J., Kralovánsky, U. P.: Biotechnology in Hungary. Vol. 69, p. 151

Honda, H., Liu, C., Kobayashi, T.: Large-Scale Plant Micropropagation. Vol. 72, p. 157

Hórvath, C. see Freitag, R.: Vol. 53, p. 17

Hou, A. see Drmanac, R.: Vol. 77, p. 75

Hubbell, E. see Hannenhalli, S.: Vol. 77, p. 1

Hummel, W.: New Alcohol Dehydrogenases for the Synthesis of Chiral Compounds. Vol. 58, p. 145

Imamoglu, S.: Simulated Moving Bed Chromatography (SMB) for Application in Bio-separation. Vol. 76, p. 211

Inada, Y., Matsushima, A., Hiroto, M., Nishimura, H., Kodera, Y.: Chemical Modifications of Proteins with Polyethylen Glycols. Vol. 52, p. 129
Irwin, D. C. see Wilson, D. B.: Vol. 65, p. 1
Isermann, H. P. see Bungay, H. R.: Vol. 70, p. 109
Iyer, P. see Lee, Y. Y.: Vol. 65, p. 93

James, E., Lee, J. M.: The Production of Foreign Proteins from Genetically Modified Plant Cells. Vol. 72, p. 127
Jeffries, T. W., Shi, N.-Q.: Genetic Engineering for Improved Xylose Fementation by Yeasts. Vol. 65, p. 117
Jendrossek, D.: Microbial Degradation of Polyesters. Vol. 71, p. 293
Jin, H. see Drmanac, R.: Vol. 77, p. 75
Jin, P. see Drmanac, R.: Vol. 77, p. 75
Johnson, D. V., Griengl, H.: Biocatalytic Applications of Hydroxynitrile. Vol. 63, p. 31
Johnson, E. A., Schroeder, W. A.: Microbial Carotenoids. Vol. 53, p. 119
Johnsurd, S. C.: Biotechnolgy for Solving Slime Problems in the Pulp and Paper Industry. Vol. 57, p. 311
Jönsson, L. J. see Hahn-Hägerdal, B.: Vol. 73, p. 53
Joshi, J. B. see Elias, C. B.: Vol. 59, p. 47
Jurinke, C., van den Boom, D., Cantor, C. R., Köster, H.: The Use of MassARRAY Technology for High Throughput Genotyping. Vol. 77, p. 57

Kaderbhai, N. see Shaw, A. D.: Vol. 66, p. 83
Karanth, N. G. see Krishna, S. H.: Vol. 75, p. 119
Kataoka, M. see Shimizu, S.: Vol. 58, p. 45
Kataoka, M. see Shimizu, S.: Vol. 63, p. 109
Katzen, R., Tsao, G. T.: A View of the History of Biochemical Engineering. Vol. 70, p. 77
Kawai, F.: Breakdown of Plastics and Polymers by Microorganisms. Vol. 52, p. 151
Kell, D. B. see Shaw, A. D.: Vol. 66, p. 83
Kessler, B., Weusthuis, R., Witholt, B., Eggink, G.: Production of Microbial Polyesters: Fermentation and Downstream Processes. Vol. 71, p. 159
Khosla, C. see McDaniel, R.: Vol. 73, p. 31
Kieran, P. M., Malone, D. M., MacLoughlin, P. F.: Effects of Hydrodynamic and Interfacial Forces on Plant Cell Suspension Systems. Vol. 67, p. 139
Kijne, J. W. see Memelink, J.: Vol. 72, p. 103
Kim, D.-I. see Choi, J. W.: Vol. 72, p. 63
Kim, Y. B., Lenz, R. W.: Polyesters from Microorganisms. Vol. 71, p. 51
King, R.: Mathematical Modelling of the Morphology of Streptomyces Species. Vol. 60, p. 95
Kino-oka, M., Nagatome, H., Taya, M.: Characterization and Application of Plant Hairy Roots Endowed with Photosynthetic Functions. Vol. 72, p. 183
Kirk, T. K. see Akhtar, M.: Vol. 57, p. 159
Knorre, W. A. see Bocker, H.: Vol. 70, p. 35
Kobayashi, M. see Shimizu, S.: Vol. 58, p. 45
Kobayashi, S., Uyama, H.: In vitro Biosynthesis of Polyesters. Vol. 71, p. 241
Kobayashi, T. see Honda, H.: Vol. 72, p. 157
Kodera, F. see Inada, Y.: Vol. 52, p. 129
Kolattukudy, P. E.: Polyesters in Higher Plants. Vol. 71, p. 1
König, A. see Riedel, K: Vol. 75, p. 81
de Koning, G. J. M. see van der Walle, G. A. M.: Vol. 71, p. 263
Konthur, Z. see Eickhoff, H.: Vol. 77, p. 103
Kossen, N. W. F.: The Morphology of Filamentous Fungi. Vol. 70, p. 1
Köster, H. see Jurinke, C.: Vol. 77, p. 57
Krabben, P., Nielsen, J.: Modeling the Mycelium Morphology of Penicilium Species in Submerged Cultures. Vol. 60, p. 125
Kralovánszky, U. P. see Holló, J.: Vol. 69, p. 151

Krämer, R.: Analysis and Modeling of Substrate Uptake and Product Release by Procaryotic and Eucaryotik Cells. Vol. 54, p. 31

Kretzmer, G.: Influence of Stress on Adherent Cells. Vol. 67, p. 123

Krieger, N. see Mitchell, D. A.: Vol. 68, p. 61

Krishna, S. H., Srinivas, N. D., Raghavarao, K. S. M. S., Karanth, N. G.: Reverse Micellar Extraction for Downstream Processeing of Proteins/Enzymes. Vol. 75, p. 119

Kuhad, R. C., Singh, A., Eriksson, K.-E. L.: Microorganisms and Enzymes Involved in the Degradation of Plant Cell Walls. Vol. 57, p. 45

Kuhad, R. Ch. see Singh, A.: Vol. 51, p. 47

Kumagai, H.: Microbial Production of Amino Acids in Japan. Vol. 69, p. 71

Kumar, S. see Harvey, N. L.: Vol. 62, p. 107

Kunze, G. see Riedel, K.: Vol. 75, p. 81

Kwon, S. see Drmanac, R.: Vol. 77, p. 75

Lacy, S. see Drmanac, R.: Vol. 77, p. 75

Ladenstein, R., Antranikian, G.: Proteins from Hyperthermophiles: Stability and Enzamatic Catalysis Close to the Boiling Point of Water. Vol. 61, p. 37

Ladisch, C. M. see Mosier, N. S.: Vol. 65, p. 23

Ladisch, M. R. see Mosier, N. S.: Vol. 65, p. 23

LaFayette, P. R. see Dean, J. F. D.: Vol. 57, p. 1

Lammers, F., Scheper, T.: Thermal Biosensors in Biotechnology. Vol. 64, p. 35

Larroche, C., Gros, J.-B.: Special Transformation Processes Using Fungal Spares and Immobilized Cells. Vol. 55, p. 179

Lazarus, M. see Adam, W.: Vol. 63, p. 73

Leak, D. J. see van der Werf, M. J.: Vol. 55, p. 147

Lee, J. M. see James, E.: Vol. 72, p. 127

Lee, S. Y., Chang, H. N.: Production of Poly(hydroxyalkanoic Acid). Vol. 52, p. 27

Lee, S. Y., Choi, J.: Production of Microbial Polyester by Fermentation of Recombinant Microorganisms. Vol. 71, p. 183

Lee, Y. Y., Iyer, P., Torget, R. W.: Dilute-Acid Hydrolysis of Lignocellulosic Biomass. Vol. 65, p. 93

Lehrach, H. see Eickhoff, H.: Vol. 77, p. 103

Lenz, R. W. see Kim, Y. B.: Vol. 71, p. 51

Licari, P. see McDaniel, R.: Vol. 73, p. 31

Lievense, L. C., van't Riet, K.: Convective Drying of Bacteria II. Factors Influencing Survival. Vol. 51, p. 71

Linden, J. C., Haigh, J. R., Mirjalili, N., Phisaphalong, M.: Gas Concentration Effects on Secondary Metabolite Production by Plant Cell Cultures. Vol. 72, p. 27

Lipshutz, R. see Hannenhalli, S.: Vol. 77, p. 1

Little, D. see Drmanac, R.: Vol. 77, p. 75

Liu, C. see Honda, H.: Vol. 72, p. 157

Lueking, A. see Eickhoff, H.: Vol. 77, p. 103

MacLoughlin, P. F. see Kieran, P. M.: Vol. 67, p. 139

Malone, D. M. see Kieran, P. M.: Vol. 67, p. 139

Maloney, S. see Müller, R.: Vol. 61, p. 155

Mandenius, C.-F.: Electronic Noses for Bioreactor Monitoring. Vol. 66, p. 65

Markov, S. A., Bazin, M. J., Hall, D. O.: The Potential of Using Cyanobacteria in Photobioreactors for Hydrogen Production. Vol. 52, p. 59

Marteinsson, V. T. see Prieur, D.: Vol. 61, p. 23

Mathur, R. K. see Ghosh, A. C.: Vol. 56, p. 111

Matsushima, A. see Inada, Y.: Vol. 52, p. 129

McCaskill, D., Croteau, R.: Prospects for the Bioengineering of Isoprenoid Biosynthesis. Vol. 55, p. 107

McDaniel, R., Licari, P., Khosla, C.: Process Development and Metabolic Engineering for the Overproduction of Natural and Unnatural Polyketides. Vol. 73, p. 31

McDonell, T. J. see Bruckheimer, E. M.: Vol. 62, p. 75

McGall, G.H., Christians, F.C.: High-Density GeneChip Oligonucleotide Probe Arrays. Vol. 77, p. 21

McGovern, A. see Shaw, A. D.: Vol. 66, p. 83

McGowan, A. J. see McKenna, S. L.: Vol. 62, p. 1

McIntyre, M., Müller, C., Dynesen, J., Nielsen, J.: Metabolic Engineering of the *Aspergillus*. Vol. 73, p. 103

McKenna, S. L., McGowan, A. J., Cotter, T. G.: Molecular Mechanisms of Programmed Cell Death. Vol. 62, p. 1

McLoughlin, A. J.: Controlled Release of Immobilized Cells as a Strategy to Regulate Ecological Competence of Inocula. Vol. 51, p. 1

Memelink, J., Kijne, J. W., van der Heijden, R., Verpoorte, R.: Genetic Modification of Plant Secondary Metabolite Pathways Using Transcriptional Regulators. Vol. 72, p. 103

Menachem, S. B. see Argyropoulos, D. S.: Vol. 57, p. 127

Menawat, A. S. see Gomes J.: Vol. 59, p. 1

Menge, M. see Mukerjee, J.: Vol. 68, p. 1

Merkle, S. A. see Dean, J. F. D.: Vol. 57, p. 1

Mirjalili, N. see Linden, J. C.: Vol. 72, p. 27

Mitchell, D. A., Berovic, M., Krieger, N.: Biochemical Engineering Aspects of Solid State Bio-processing. Vol. 68, p. 61

Moeur, B. see Drmanac, R.: Vol. 77, p. 75

Moore, J. C. see Arnold, F. H.: Vol. 58, p. 1

Moracci, M. see van der Oost, J.: Vol. 61, p. 87

Mosier, N. S., Hall, P., Ladisch, C. M., Ladisch, M. R.: Reaction Kinetics, Molecular Action, and Mechanisms of Cellulolytic Proteins. Vol. 65, p. 23

Mühlemann, H. M., Bungay, H. R.: Research Perspectives for Bioconversion of Scrap Paper. Vol. 65, p. 193

Mukherjee, J., Menge, M.: Progress and Prospects of Ergot Alkaloid Research. Vol. 68, p. 1

Mukhopadhyay, A.: Inclusion Bodies and Purification of Proteins in Biologically Active Forms. Vol. 56, p. 61

Mukhopadhyay, A. see Bhatia, P. K.: Vol. 64, p. 155

Müller, C. see McIntyre, M.: Vol. 73, p. 103

Müller, R., Antranikian, G., Maloney, S., Sharp, R.: Thermophilic Degradation of Environmental Pollutants. Vol. 61, p. 155

Nagatome, H. see Kino-oka, M.: Vol. 72, p. 183

Nagy, E.: Three-Phase Oxygen Absorption and its Effect on Fermentation. Vol. 75, p. 51

Necina, R. see Strancar, A.: Vol. 76, p. 49

Nielsen, J. see Christensen, B.: Vol. 66, p. 209

Nielsen, J. see Krabben, P.: Vol. 60, p. 125

Nielsen, J. see McIntyre, M.: Vol. 73, p. 103

Nisbet, L. J. see Hill, D. C.: Vol. 59, p. 73

Nishimura, H. see Inada, Y.: Vol. 52, p. 123

Nordhoff, E. see Eickhoff, H.: Vol. 77, p. 103

Nyarsik, L. see Eickhoff, H.: Vol. 77, p. 103

Ochsner, U. A., Hembach, T., Fiechter, A.: Produktion of Rhamnolipid Biosurfactants. Vol. 53, p. 89

O'Connor, R.: Survival Factors and Apoptosis: Vol. 62, p. 137

Ogawa, J. see Shimizu, S.: Vol. 58, p. 45

Ohta, H.: Biocatalytic Asymmetric Decarboxylation. Vol. 63, p. 1

Oliverio, S. see Autuori, F.: Vol. 62, p. 129

van der Oost, J., Ciaramella, M., Moracci, M., Pisani, F.M., Rossi, M., de Vos, W.M.: Molecular Biology of Hyperthermophilic Archaea. Vol. 61, p. 87

Orlich, B., Schomäcker, R.: Enzyme Catalysis in Reverse Micelles. Vol. 75, p. 185
Orru, R. V. A., Archelas, A., Furstoss, R., Faber, K.: Epoxide Hydrolases and Their Synthetic Applications. Vol. 63, p. 145
Osbourn, A. E. see Haralampidis, D.: Vol. 75, p. 31

Paul, G. C., Thomas, C. R.: Characterisation of Mycelial Morphology Using Image Analysis. Vol. 60, p. 1
Perrier, M. see Dochain, D.: Vol. 56, p. 147
Pevzner, P. A. see Hannenhalli, S.: Vol. 77, p. 1
Phisaphalong, M. see Linden, J. C.: Vol. 72, p. 27
Piacentini, G. see Autuori, F.: Vol. 62, p. 129
Piredda, L. see Autuori, F.: Vol. 62, p. 129
Pisani, F. M. see van der Oost, J.: Vol. 61, p. 87
Podgornik, A. see Strancar, A.: Vol. 76, p. 49
Podgornik, A,. Tennikova, T. B.: Chromatographic Reactors Based on Biological Activity. Vol. 76, p. 165
Pohl, M.: Protein Design on Pyruvate Decarboxylase (PDC) by Site-Directed Mutagenesis. Vol. 58, p. 15
Poirier, Y.: Production of Polyesters in Transgenic Plants. Vol. 71, p. 209
Pons, M.-N., Vivier, H.: Beyond Filamentous Species. Vol. 60, p. 61
Pons, M.-N., Vivier, H.: Biomass Quantification by Image Analysis. Vol. 66, p. 133
Prieur, D., Marteinsson, V. T.: Prokaryotes Living Under Elevated Hydrostatic Pressure. Vol. 61, p. 23
Prior, A. see Wolfgang, J.: Vol. 76, p. 233
Pulz, O., Scheibenbogen, K.: Photobioreactors: Design and Performance with Respect to Light Energy Input. Vol. 59, p. 123

Raghavarao, K. S. M. S., Dueser, M., Todd, P.: Multistage Magnetic and Electrophoretic Extraction of Cells, Particles and Macromolecules. Vol. 68, p. 139
Raghavarao, K. S. M. S. see Krishna, S. H.: Vol. 75, p. 119
Ramanathan, K. see Xie, B.: Vol. 64, p. 1
Riedel, K., Kunze, G., König, A.: Microbial Sensor on a Respiratory Basis for Wastewater Monitoring. Vol. 75, p. 81
van't Riet, K. see Lievense, L. C.: Vol. 51, p. 71
Roberts, S. M. see Allan, J. V.: Vol. 63, p. 125
Robinson, A. see Brazma, A.: Vol. 77, p. 113
Roehr, M.: History of Biotechnology in Austria. Vol. 69, p. 125
Rogers, P. L., Shin, H. S., Wang, B.: Biotransformation for L-Ephedrine Production. Vol. 56, p. 33
Rossi, M. see van der Oost, J.: Vol. 61, p. 87
Rowland, J. J. see Shaw, A. D.: Vol. 66, p. 83
Roychoudhury, P. K., Srivastava, A., Sahai, V.: Extractive Bioconversion of Lactic Acid. Vol. 53, p. 61
Rusin, P., Ehrlich, H. L.: Developments in Microbial Leaching – Mechanisms of Manganese Solubilization. Vol. 52, p. 1
Russell, N. J.: Molecular Adaptations in Psychrophilic Bacteria: Potential for Biotechnological Applications. Vol. 61, p. 1

Sablon, E., Contreras, B., Vandamme, E.: Antimicrobial Peptides of Lactic Acid Bacteria: Mode of Action, Genetics and Biosynthesis. Vol. 68, p. 21
Sahai, V. see Singh, A.: Vol. 51, p. 47
Sahai, V. see Roychoudhury, P. K.: Vol. 53, p. 61
Saha-Möller, C. R. see Adam, W.: Vol. 63, p. 73
Sahm, H. see Eggeling, L.: Vol. 54, p. 1
Sahm, H. see de Graaf, A. A.: Vol. 73, p. 9

Sahoo, G. C., Dutta, N. N.: Perspectives in Liquid Membrane Extraction of Cephalosporin Antibiotics: Vol. 75, p. 209

Saleemuddin, M.: Bioaffinity Based Immobilization of Enzymes. Vol. 64, p. 203

Santos, H. see da Costa, M. S.: Vol. 61, p. 117

Sarkans, U. see Brazma, A.: Vol. 77, p. 113

Sarkiss, M. see Bruckheimer, E. M.: Vol. 62, p. 75

Sauer, U.: Evolutionary Engineering of Industrially Important Microbial Phenotypes. Vol. 73, p. 129

Scheibenbogen, K. see Pulz, O.: Vol. 59, p. 123

Scheper, T. see Lammers, F.: Vol. 64, p. 35

Schneider, K. see Beyeler, W.: Vol. 70, p. 139

Schomäcker, R. see Orlich, B.: Vol. 75, p. 185

Schreier, P.: Enzymes and Flavour Biotechnology. Vol. 55, p. 51

Schreier, P. see Adam, W.: Vol. 63, p. 73

Schroeder, W. A. see Johnson, E. A.: Vol. 53, p. 119

Schügerl, K., Gerlach, S. R., Siedenberg, D.: Influence of the Process Parameters on the Morphology and Enzyme Production of Aspergilli. Vol. 60, p. 195

Schügerl, K. see Seidel, G.: Vol. 66, p. 115

Schügerl, K.: Recovery of Proteins and Microorganisms from Cultivation Media by Foam Flotation. Vol. 68, p. 191

Schügerl, K.: Development of Bioreaction Engineering. Vol. 70, p. 41

Schumann, W.: Function and Regulation of Temperature-Inducible Bacterial Proteins on the Cellular Metabolism. Vol. 67, p. 1

Schuster, K. C.: Monitoring the Physiological Status in Bioprocesses on the Cellular Level. Vol. 66, p. 185

Scouroumounis, G. K. see Winterhalter, P.: Vol. 55, p. 73

Scragg, A.H.: The Production of Aromas by Plant Cell Cultures. Vol. 55, p. 239

Sedlak, M. see Ho, N. W. Y.: Vol. 65, p. 163

Seidel, G., Tollnick, C., Beyer, M., Schügerl, K.: On-line and Off-line Monitoring of the Production of Cephalosporin C by Acremonium Chrysogenum. Vol. 66, p. 115

Shafto, J. see Drmanac, R.: Vol. 77, p. 75

Shamlou, P. A. see Yim, S. S.: Vol. 67, p. 83

Shapira, M. see Gutman, A. L.: Vol. 52, p. 87

Sharp, R. see Müller, R.: Vol. 61, p. 155

Shaw, A. D., Winson, M. K., Woodward, A. M., McGovern, A., Davey, H. M., Kaderbhai, N., Broadhurst, D., Gilbert, R. J., Taylor, J., Timmins, E. M., Alsberg, B. K., Rowland, J. J., Goodacre, R., Kell, D. B.: Rapid Analysis of High-Dimensional Bioprocesses Using Multivariate Spectroscopies and Advanced Chemometrics. Vol. 66, p. 83

Shi, N.-Q. see Jeffries, T. W.: Vol. 65, p. 117

Shimizu, S. see Hasegawa, S.: Vol. 51, p. 91

Shimizu, S., Ogawa, J., Kataoka, M., Kobayashi, M.: Screening of Novel Microbial for the Enzymes Production of Biologically and Chemically Useful Compounds. Vol. 58, p. 45

Shimizu, S., Kataoka, M.: Production of Chiral C3- and C4-Units by Microbial Enzymes. Vol. 63, p. 109

Shin, H. S. see Rogers, P. L.: Vol. 56, p. 33

Siedenberg, D. see Schügerl, K.: Vol. 60, p. 195

Singh, A., Kuhad, R. Ch., Sahai, V., Ghosh, P.: Evaluation of Biomass. Vol. 51, p. 47

Singh, A. see Kuhad, R. C.: Vol. 57, p. 45

Singh, R. P., Al-Rubeai, M.: Apoptosis and Bioprocess Technology. Vol. 62, p. 167

Sohail, M., Southern, E. M.: Oligonucleotide Scanning Arrays: Application to High-Through-put Screening for Effective Antisense Reagents and the Study of Nucleic Acid Inter-actions. Vol. 77, p. 43

Sonnleitner, B.: New Concepts for Quantitative Bioprocess Research and Development. Vol. 54, p. 155

Sonnleitner, B.: Instrumentation of Biotechnological Processes. Vol. 66, p. 1

Southern, E. M. see Sohail, M.: Vol. 77, p. 43

Srinivas, N. D. see Krishna, S. H.: Vol. 75, p. 119

Srivastava, A. see Roychoudhury, P. K.: Vol. 53, p. 61

Stafford, D. E., Yanagimachi, K. S., Stephanopoulos, G.: Metabolic Engineering of Indene Bioconversion in *Rhodococcus sp.* Vol. 73, p. 85

Stefuca, V., Gemeiner, P.: Investigation of Catalytic Properties of Immobilized Enzymes and Cells by Flow Microcalorimetry. Vol. 64, p. 69

Steinbüchel, A., Hein, S.: Biochemical and Molecular Basis of Microbial Synthesis of Poly-hydroxyalkanoates in Microorganisms. Vol. 71, p. 81

Stephanopoulos, G., Gill, R. T.: After a Decade of Progress, an Expanded Role for Metabolic Engineering. Vol. 73, p. 1

Stephanopoulos, G. see Stafford, D. E.: Vol. 73, p. 85

Strancar, A., Podgornik, A., Barut, M., Necina, R.: Short Monolithic Columns as Stationary Phases for Biochromatography. Vol. 76, p. 49

Suurnäkki, A., Tenkanen, M., Buchert, J., Viikari, L.: Hemicellulases in the Bleaching of Chemical Pulp. Vol. 57, p. 261

Svec, F.: Capillary Electrochromatography: a Rapidly Emerging Separation Method. Vol. 76, p. 1

Svec, F. see Xie, S.: Vol. 76, p. 87

Swanson, D. see Drmanac, R.: Vol. 77, p. 75

Taya, M. see Kino-oka, M.: Vol. 72, p. 183

Taylor, J. see Shaw, A. D.: Vol. 66, p. 83

Tenkanen, M. see Suurnäkki, A.: Vol. 57, p. 261

Tennikova, T. B. see Podgornik, A.: Vol. 76, p. 165

Thiericke, R. see Grabely, S.: Vol. 64, p. 101

Thomas, C. R. see Paul, G. C.: Vol. 60, p. 1

Thömmes, J.: Fluidized Bed Adsorption as a Primary Recovery Step in Protein Purification. Vol. 58, p. 185

Timmens, E. M. see Shaw, A. D.: Vol. 66, p. 83

Todd, P. see Raghavarao, K. S. M. S.: Vol. 68, p. 139

Tolan, J. S., Guenette, M.: Using Enzymes in Pulp Bleaching: Mill Applications. Vol. 57, p. 289

Tolan, J. S., Foody, B.: Cellulase from Submerged Fermentation. Vol. 65, p. 41

Tollnick, C. see Seidel, G.: Vol. 66, p. 115

Torget, R. W. see Lee, Y. Y.: Vol. 65, p. 93

Traganos, F. see Darzynkiewicz, Z.: Vol. 62, p. 33

Trojanowska, M. see Haralampidis, D.: Vol. 75, p. 31

Tsao, G. T., Cao, N. J., Du, J., Gong, C. S.: Production of Multifunctional Organic Acids from Renewable Resources. Vol. 65, p. 243

Tsao, G. T. see Gong, C. S.: Vol. 65, p. 207

Tsao, G. T. see Katzen, R.: Vol. 70, p. 77

Ukrainczyk, T. see Drmanac, R.: Vol. 77, p. 75

Uyama, H. see Kobayashi, S.: Vol. 71, p. 241

Vandamme, E. see Sablon, E.: Vol. 68, p. 21

Verpoorte, R. see Memelink, J.: Vol. 72, p. 103

Viikari, L. see Suurnäkki, A.: Vol. 57, p. 261

Vilo, J. see Brazma, A.: Vol. 77, p. 113

Vingron, M. see Brazma, A.: Vol. 77, p. 113

Vivier, H. see Pons, M.-N.: Vol. 60, p. 61

Vivier, H. see Pons, M.-N.: Vol. 66, p. 133

de Vos, W. M. see van der Oost, J.: Vol. 61, p. 87

Wahlbom, C. F. see Hahn-Hägerdal, B.: Vol. 73, p. 53

Wall, M. B. see Farrell, R. L.: Vol. 57, p. 197

van der Walle, G. A. M., de Koning, G. J. M., Weusthuis, R. A., Eggink, G.: Properties, Modifications and Applications of Biopolyester. Vol. 71, p. 263

Walter, G. see Eickhoff, H.: Vol. 77, p. 103

Wang, B. see Rogers, P. L.: Vol. 56, p. 33

Weichold, O. see Adam, W.: Vol. 63, p. 73

van der Werf, M. J., de Bont, J. A. M. Leak, D. J.: Opportunities in Microbial Biotransformation of Monoterpenes. Vol. 55, p. 147

Weuster-Botz, D., de Graaf, A.A.: Reaction Engineering Methods to Study Intracellular Metabolite Concentrations. Vol. 54, p. 75

Weusthuis, R. see Kessler, B.: Vol. 71, p. 159

Weusthuis, R. A. see van der Walle, G. J. M.: Vol. 71, p. 263

Wiechert, W., de Graaf, A.A.: In Vivo Stationary Flux Analysis by ^{13}C-Labeling Experiments. Vol. 54, p. 109

Wiesmann, U.: Biological Nitrogen Removal from Wastewater. Vol. 51, p. 113

Williamson, N. M. see Allan, J. V.: Vol. 63, p. 125

Wilson, D. B., Irwin, D. C.: Genetics and Properties of Cellulases. Vol. 65, p. 1

Winson, M. K. see Shaw, A. D.: Vol. 66, p. 83

Winterhalter, P., Skouroumounis, G. K.: Glycoconjugated Aroma Compounds: Occurence, Role and Biotechnological Transformation. Vol. 55, p. 73

Witholt, B. see Kessler, B.: Vol. 71, p. 159

Wolfgang, J., Prior, A.: Continuous Annular Chromatography. Vol. 76, p. 233

Woodley, J. M.: Advances in Enzyme Technology – UK Contributions. Vol. 70, p. 93

Woodward, A. M. see Shaw, A. D.: Vol. 66, p. 83

Wrigley, S. K. see Hill, D. C.: Vol. 59, p. 73

Xia, L. see Cen, P.: Vol. 65, p. 69

Xie, B., Ramanathan, K., Danielsson, B.: Principles of Enzyme Thermistor Systems: Applications to Biomedical and Other Measurements. Vol. 64, p. 1

Xie, S., Allington, R. W., Fréchet, J. M. J., Svec, F.: Porous Polymer Monoliths: An Alternative to Classical Beads. Vol. 76, p. 87

Xu, C. see Drmanac, R.: Vol. 77, p. 75

Yanagimachi, K. S. see Stafford, D. E.: Vol. 73, p. 85

Yim, S. S., Shamlou, P. A.: The Engineering Effects of Fluids Flow and Freely Suspended Biological Macro-Materials and Macromolecules. Vol. 67, p. 83

Zhong, J.-J.: Biochemical Engineering of the Production of Plant-Specific Secondary Metabolites by Cell Suspension Cultures. Vol. 72, p. 1

van Zyl, W. H. see Hahn-Hägerdal, B.: Vol. 73, p. 53

Subject Index

ABI DNA/RNA synthesiser 48
ABO blood groups 65
Abrasive paper 48
N-Acetyltransferase isoenzymes 66
Affymetrix 21
Aggregation 122
Aliphatic chain 47
Ammonia 50
Anhydrous ammonia 48
Antibodies 108
–, autoantibodies 109
Antibody profiling 103
Antisense oligonucleotides 40
Antisense reagents 53
ApoB 94
Array hybridization 123, 131
Array scanning 127
Arrays, DNA 21
–, micromirror 2
–, uniform 2
Assay arrays 2
Association 122
– cardinality 122
– role 122
Autoimmune diseases 108, 109
Automation 103

Bacteriophage PhiX174 93
Bacteriophage promoter 51
Bases, degenerate 83
Beam technologies 18
Beaucage reagent 50
Bioevent 133
Biomaterial 132
Biosequence 125
Border Minimization Problem (BMP) 2, 18
Branching points 81
Bridge genomics 103

Cancer profiling, classification 33–34
Capping reaction 24, 27
Carbohydrates, MALDI TOF 61
Carbonate/bicarbonate buffer 51

cDNA clones 104
cDNA microarrays 105
Cell-free egg extracts 54
Circular-mask 44
Class attribute 122
Cleavable linker 50
Clone libraries 104
Clones, overlapping 81
Clover-leaf 55
Combinatorial algorithms 1
Controlled pore glass (CPG) 50
Coupling reaction 24, 27
Crocus paper 48
Cyclins 54
Cytochrome P450 66

Deprotection 50
Diamond-shaped mask 44
Dideoxy gel sequencing method 85
Dideoxynucleotides (ddNTPs) 67
Digital light processors 30
Direct visualization 76
Directed evolution 38–39
DNA 104
– arrays 1, 2, 21, 85
– chip 57
– hybridisation 104
–, junk 63
–, MALDI 62
DNA ligase 89, 90
DNA ligation 75
DNA MassARRAY 69
DNA microarrays 108
DNA polymerase 89
DNA sequencing, SBH 75
–, de novo 85

Electron impact 58
Electrospray ionization 58
Enzymatic manipulation 50
Epoxide group 47
ESI MS 58
Ethanolamine 51

N-4-Ethyldeoxycytidine   55
Exocyclic amines   50
Exonuclease sequencing   76
Expression monitoring   75
Expression profiling   103
Expression value dimensions   134
Extract   126, 132

G:C base pair   55
GAPDH   106
Gel electrophoresis   105
Gel pads   107
Gel sequencing   75
Gene expression   104
– – database   115, 136
– – levels   115
– – matrix   115, 134
– – monitoring arrays   31–34
GeneChip DNA arrays   21
Genes   104
Genetic diversity detection   66
GenFlex arrays   37–39
Genomic databases   104
Genomics   57, 103
Genotyping arrays   34–38
–, high throughput   57, 67
–, HyChip   93
–, SBH   75
Global minimum free energy state   54
Glycidoxypropyl trimethoxysilane   47
Glyco-conjugates, MALDI TOF   61
Glycopeptides, MALDI TOF   61
Glycoproteins, MALDI TOF   61
Gray codes   2

Heteroduplex formation   53, 55
Heterozygotes, HyChip   93, 94
High throughput genotyping   57
HIV   91–93
Homo sapiens   63
Human Genome Project   63
Hybridization   51
– chemistry   97
–, competitive   81
–, conditions   83
–, contiguous stacking   81
– efficiency   83
– probes   78
– specificity   83
HyChip sequencing chip   91–94
Hydroxyl group   47

Immobilisation, proteins   107
Immobilised metal affinity chromatography
   (IMAC)   106

In vitro transcription   51
In vitro translation   54
Intermolecular binding   53
Intramolecular folding   53
Ionization   58

Labeling   127, 133
Leader sequence   51
Ligase   89, 90
Ligation   89, 91
–, combinatorial   91
Light-directed synthesis   21
Linker, cleavable disulfide   27
Linker/spacer   47

MALDI MS   57, 59, 109
Mask   2
– decomposition problem   14, 18
– design   1
Mass spectrometry   57
MassARRAY   67
MassEXTEND   66
Matching problem   6
Microarray data, derived   119, 128
– –, raw   119, 128
Microarray design   124
Microarray element   124, 131
Microarray experiment   123, 131
– – annotation   116, 138
– – steps   117
Microarray images   137
– – analysis   127, 134
Microarrays, biomalecules   105
–, molecular interactions   108
Microfluidic chips   107
Micromirror arrays   2
Microsatellites   64
Microtitre plates   105
Minimum length cycle cover problem   5
Modified bases   55
Modified nucleotides   50
Molecular barcode   38
Mutation/SNP discovery   75
Mutation detection, SBH   78
– – arrays   34–38
Mutation discovery   86

Nano-barcode tags   96
Nanotags   96
Non-array methods   95
Nucleic acid folding   54
Nucleic acids, locked   97
– –, MALDI MS   62
Nucleoside phoshoramidite   24
Nucleotide CE phosphoramidite   47

Object class 122
Object modeling 122
Oligonucleotide microarrays 105
Oligonucleotide overlap principle 79
Oligonucleotide probe arrays 27
Oligonucleotides 58, 75, 87
–, antisense 40
Ontology 116, 130
Oocytes 54
Optimal placement problem 18
Optimal threading problem 8
Oxidation 24, 27
Oxidising agent 50

Parallel data collection, SBH 84
Pattern generator technology 18
PCR, MALDI 63, 66
Peptide nucleic acids 97
Peptides, MALDI TOF MS 61
Phage display antibody libraries 108
Phagemid 108
Pharmacogenetics 65
Pharmacogenomics 65
Phosphoramidite 24
PhosphorImager 52
Phosphorothioate bond 50
Photo-generated acid catalyst 30
Photochemical deprotection 23–31
Photolabile protecting group 24, 26, 29
Photolithography 1, 23–31
Placement problem 3, 18
Plasma desorption 58
Plasma discharge 48
Polyacrylamide gel pad microarrays 107
Polyethylene glycol 47
Polymerase chain reaction (PCR) 51
Polymorphic markers 63
Polymorphism 67
– detection arrays 34–38
– discovery 86
Polypropylene 47
Polyvinylidene difluoride (PVDF) 106
Pressure moulding 48
Probe, degenerate 83
Probe arrays 87
Probe design, SBH 83
Probe length 79
– –, SBH 81
Probe overlaps 80
Probe sets, universal/non-universal, SBH
  82, 88
Probes, SBH 79
Protein arrays 103
– –, living 110
– –, requirements 106

Protein deposition 109
Protein filter array technology 106
Protein microarrays 105
Protein purification 103
Protein screening arrays 38–39
Proteins, MALDI TOF MS 61
–, planar immobilisation 107
Proteomics 103
PTFE (Teflon) 48
Pyrosequencing 75

Rectangle cover problem 1, 3, 12
Relationships, subclass/aggregate/associa-
  tion 122
Reporter 124, 132
Resequencing 75
Resolution, photolithographic 26, 29–30
Restriction endonuclease 51
Reverse phosphoramidites 50
RFLPs 64
Ribozymes 53
RNA 104
–, MALDI 62
tRNA^{phe} 55
RNA folding 54
RNA polymerase 51
RNase H 53

Saccharomyces cerevisiae 110
Sample, extract preparation 126, 132
Sample source, extract preparation 125,
  132
Sample treatment, extract 126, 133
Sanger method 76
SBH 75 ff
– blind test 90
Scanning arrays 44
Sequence variations 63
Sequencing 104
–, combinatorial 89
–, de novo 75
–, direct 76
–, indirect 77
Sequencing by Hybridization (SBH) 75 ff
– – –, history 77, 78
– – –, probe length 79
Sequencing chip 91
Serum screening 103
Short tandem repeats 64
Silanation 24
Silanol groups 47
Single-base extesnion assay 37
Single-nucleotide polymorphisms 57, 64
SNP analysis arrays 35–37
SNPs 57, 64

Solid-phase synthesis   50
SpectroREAD   70
Splice variant detection   40
Storage phosphor screen   52
STR markers   64
Subclass relationship   122
Sulfurising agent   50
Synthetic pathway   54

Tag array   37 – 39
Target length   81
Template grid   53
Tetramethyl ammonium chloride   83
Theoretical free energy calculations   55
Threading   3, 6, 18

–, optimal threading problem   8
Time-of-flight (TOF)   57, 59
Transcriptional mapping   40
Transformation   136
Traveling Salesman Problem (TSP)   2, 18
Trityl yield   50

Universal arrays   94
Universal protein array system   108

V LSIPS   2
Vacuum furnace   48

Xenopus laevis   54
xvseq   52